"十三五"应用型人才培养规划教材

JSP 网站设计

陈恒 楼偶俊 主编 朱毅 项聪 张术梅 副主编

清华大学出版社
北京

内 容 简 介

本书采用"教学做"一体化模式编写,合理地组织学习单元,并将每个单元分解为核心知识、能力目标、任务驱动、实践环节4个模块。全书共包括11章,包括Web前端基础、JSP简介及开发环境的构建、JSP语法、JSP内置对象、JSP与JavaBean、JSP访问数据库、Java Servlet、过滤器、EL与JSTL、文件的上传与下载以及地址簿管理信息系统。书中实例侧重实用性和启发性,趣味性强,通俗易懂,使读者能够快速掌握Java Web应用的基础知识、编程技巧以及完整的开发体系,为适应实战应用打下坚实的基础。

本书可以作为高等职业院校、本科院校计算机及相关专业的教材或教学参考书,也适合作为Java Web开发人员的参考用书。

本书封面贴有清华大学出版社防伪标签,无标签者不得销售。
版权所有,侵权必究。举报:010-62782989,beiqinquan@tup.tsinghua.edu.cn。

图书在版编目(CIP)数据

JSP网站设计/陈恒,楼偶俊主编. —北京:清华大学出版社,2017(2022.7重印)
("十三五"应用型人才培养规划教材)
ISBN 978-7-302-47282-7

Ⅰ. ①J… Ⅱ. ①陈… ②楼… Ⅲ. ①JAVA语言-网页制作工具-高等学校-教材 Ⅳ. ①TP312.8 ②TP393.092.2

中国版本图书馆CIP数据核字(2017)第122636号

责任编辑:田在儒
封面设计:王跃宇
责任校对:赵琳爽
责任印制:曹婉颖

出版发行:清华大学出版社
  网　　址:http://www.tup.com.cn,http://www.wqbook.com
  地　　址:北京清华大学学研大厦A座　　　邮　编:100084
  社 总 机:010-83470000　　　邮　购:010-62786544
  投稿与读者服务:010-62776969,c-service@tup.tsinghua.edu.cn
  质量反馈:010-62772015,zhiliang@tup.tsinghua.edu.cn
  课件下载:http://www.tup.com.cn,010-62770175-4278
印 装 者:天津鑫丰华印务有限公司
经　　销:全国新华书店
开　　本:185mm×260mm　　印　张:17.5　　字　数:398千字
版　　次:2017年8月第1版　　　　　　　　印　次:2022年7月第4次印刷
定　　价:59.00元

产品编号:073105-02

前 言
FOREWORD

尽管已经有许多针对本科生的Java Web开发教材,但大部分教材仍侧重讲授知识,而且非常注重知识的系统性,使教材知识体系结构过于全面、庞大。目前,让学生尽快掌握最有用的知识,并尽可能地挖掘他们使用这些知识解决实际问题的能力是非常重要的,一旦做到这一点,就容易培养学生自主学习的能力,相对罗列大量知识的讲授起到事半功倍的效果。许多教师在教学过程中,非常希望教材本身能引导学生尽可能地参与教学活动,因此本书的重点不是简单地介绍Java Web开发的基础知识,而是大量的实例与实践环节,读者通过本书可以快速提高Java Web应用的开发能力。

全书共11章。第1章概括地介绍Web前端的基础知识,包括HTML、CSS与JavaScript。第2章介绍JSP运行环境的构建,并通过一个简单的Web应用讲解Java Web开发的基本步骤。第3章讲述JSP语法,包括Java脚本元素以及常用的JSP标记。第4章介绍常见的JSP内置对象,包括request、response、session以及application。第5章介绍JSP与JavaBean,JSP和JavaBean技术的结合不仅可以实现数据的表示和处理分离,而且可以提高代码重用的程度。第6章详细地介绍在JSP中如何访问关系数据库,如Oracle、SQL Server、MySQL和Microsoft Access等。第7章讲述Servlet的运行原理以及基于Servlet的MVC模式,是本书的重点内容之一。第8章详细地讲述过滤器的概念、运行原理以及实际应用。过滤器可以过滤浏览器对服务器的请求,也可以过滤服务器对浏览器的响应。第9章主要介绍EL与JSTL核心标签库的基本用法。第10章重点介绍Servlet 3.0中的HttpServletRequest对文件上传的支持。第11章是本书的重点内容之一,将前面章节的知识进行全面综合,详细讲解一个基于Servlet MVC模式的地址簿管理信息系统的开发过程。

本书特别注重引导学生参与课堂教学活动,适合作为大学计算机及相关专业的教材或教学参考书,也适合作为Java Web开发人员的参考用书。

为了便于教学,本书配有教学课件、源代码以及实践环节与课后习题参考答案,读者可从清华大学出版社网站免费下载。

由于编者水平有限,书中难免存在疏漏之处,敬请广大读者给予批评指正。

编 者
2017年5月

目 录
CONTENTS

第1章 Web前端基础 ………………………………………………… 1

1.1 HTML ……………………………………………………………… 1
    1.1.1 核心知识 …………………………………………………… 1
    1.1.2 能力目标 …………………………………………………… 12
    1.1.3 任务驱动 …………………………………………………… 12
    1.1.4 实践环节 …………………………………………………… 13

1.2 CSS ………………………………………………………………… 13
    1.2.1 核心知识 …………………………………………………… 13
    1.2.2 能力目标 …………………………………………………… 24
    1.2.3 任务驱动 …………………………………………………… 24
    1.2.4 实践环节 …………………………………………………… 25

1.3 JavaScript ………………………………………………………… 28
    1.3.1 核心知识 …………………………………………………… 28
    1.3.2 能力目标 …………………………………………………… 48
    1.3.3 任务驱动 …………………………………………………… 49
    1.3.4 实践环节 …………………………………………………… 50

1.4 小结 ……………………………………………………………… 50
习题1 …………………………………………………………………… 50

第2章 JSP简介及开发环境的构建 …………………………………… 53

2.1 构建开发环境 ……………………………………………………… 53
    2.1.1 核心知识 …………………………………………………… 53
    2.1.2 能力目标 …………………………………………………… 54
    2.1.3 任务驱动 …………………………………………………… 54
    2.1.4 实践环节 …………………………………………………… 59

2.2 使用Eclipse开发Web应用 ……………………………………… 59
    2.2.1 核心知识 …………………………………………………… 59
    2.2.2 能力目标 …………………………………………………… 59
    2.2.3 任务驱动 …………………………………………………… 59
    2.2.4 实践环节 …………………………………………………… 63

2.3 小结 ……………………………………………………………… 63
习题2 …………………………………………………………………… 64

第 3 章　JSP 语法 ... 65

3.1　JSP 页面的基本构成 ... 65
- 3.1.1　核心知识 ... 65
- 3.1.2　能力目标 ... 65
- 3.1.3　任务驱动 ... 65
- 3.1.4　实践环节 ... 67

3.2　Java 程序片 ... 67
- 3.2.1　核心知识 ... 67
- 3.2.2　能力目标 ... 68
- 3.2.3　任务驱动 ... 68
- 3.2.4　实践环节 ... 69

3.3　成员变量和方法的声明 ... 70
- 3.3.1　核心知识 ... 70
- 3.3.2　能力目标 ... 70
- 3.3.3　任务驱动 ... 70
- 3.3.4　实践环节 ... 71

3.4　Java 表达式 ... 72
- 3.4.1　核心知识 ... 72
- 3.4.2　能力目标 ... 72
- 3.4.3　任务驱动 ... 72
- 3.4.4　实践环节 ... 73

3.5　page 指令标记 ... 73
- 3.5.1　核心知识 ... 73
- 3.5.2　能力目标 ... 74
- 3.5.3　任务驱动 ... 74
- 3.5.4　实践环节 ... 75

3.6　include 指令标记 ... 75
- 3.6.1　核心知识 ... 75
- 3.6.2　能力目标 ... 75
- 3.6.3　任务驱动 ... 75
- 3.6.4　实践环节 ... 77

3.7　include 动作标记 ... 77
- 3.7.1　核心知识 ... 77
- 3.7.2　能力目标 ... 77
- 3.7.3　任务驱动 ... 77
- 3.7.4　实践环节 ... 78

3.8　forward 动作标记 ... 78
- 3.8.1　核心知识 ... 78
- 3.8.2　能力目标 ... 79
- 3.8.3　任务驱动 ... 79
- 3.8.4　实践环节 ... 80

3.9　param 动作标记 ... 80
- 3.9.1　核心知识 ... 80

3.9.2 能力目标 …… 81
3.9.3 任务驱动 …… 81
3.9.4 实践环节 …… 82
3.10 小结 …… 82
习题 3 …… 83

第 4 章　JSP 内置对象 …… 84

4.1 请求对象 request …… 84
 4.1.1 核心知识 …… 84
 4.1.2 能力目标 …… 86
 4.1.3 任务驱动 …… 87
 4.1.4 实践环节 …… 90
4.2 应答对象 response …… 90
 4.2.1 核心知识 …… 90
 4.2.2 能力目标 …… 91
 4.2.3 任务驱动 …… 91
 4.2.4 实践环节 …… 95
4.3 会话对象 session …… 95
 4.3.1 核心知识 …… 95
 4.3.2 能力目标 …… 103
 4.3.3 任务驱动 …… 103
 4.3.4 实践环节 …… 105
4.4 全局应用程序对象 application …… 106
 4.4.1 核心知识 …… 106
 4.4.2 能力目标 …… 107
 4.4.3 任务驱动 …… 107
 4.4.4 实践环节 …… 109
4.5 小结 …… 109
习题 4 …… 109

第 5 章　JSP 与 JavaBean …… 111

5.1 编写 JavaBean …… 111
 5.1.1 核心知识 …… 111
 5.1.2 能力目标 …… 112
 5.1.3 任务驱动 …… 112
 5.1.4 实践环节 …… 113
5.2 JSP 中使用 JavaBean …… 113
 5.2.1 核心知识 …… 113
 5.2.2 能力目标 …… 119
 5.2.3 任务驱动 …… 119
 5.2.4 实践环节 …… 121
5.3 小结 …… 122
习题 5 …… 122

第6章　JSP 访问数据库 …… 124

6.1　使用 JDBC-ODBC 桥接器连接数据库 …… 124
- 6.1.1　核心知识 …… 124
- 6.1.2　能力目标 …… 125
- 6.1.3　任务驱动 …… 125
- 6.1.4　实践环节 …… 129

6.2　使用纯 Java 数据库驱动程序连接数据库 …… 130
- 6.2.1　核心知识 …… 130
- 6.2.2　能力目标 …… 130
- 6.2.3　任务驱动 …… 131
- 6.2.4　实践环节 …… 134

6.3　Statement、ResultSet 的使用 …… 134
- 6.3.1　核心知识 …… 134
- 6.3.2　能力目标 …… 134
- 6.3.3　任务驱动 …… 135
- 6.3.4　实践环节 …… 139

6.4　游动查询 …… 140
- 6.4.1　核心知识 …… 140
- 6.4.2　能力目标 …… 140
- 6.4.3　任务驱动 …… 140
- 6.4.4　实践环节 …… 143

6.5　访问 Excel 电子表格 …… 143
- 6.5.1　核心知识 …… 143
- 6.5.2　能力目标 …… 144
- 6.5.3　任务驱动 …… 144
- 6.5.4　实践环节 …… 146

6.6　使用连接池 …… 146
- 6.6.1　核心知识 …… 146
- 6.6.2　能力目标 …… 147
- 6.6.3　任务驱动 …… 147
- 6.6.4　实践环节 …… 150

6.7　其他典型数据库的连接 …… 150
- 6.7.1　核心知识 …… 150
- 6.7.2　能力目标 …… 151
- 6.7.3　任务驱动 …… 151
- 6.7.4　实践环节 …… 153

6.8　PreparedStatement 的使用 …… 153
- 6.8.1　核心知识 …… 153
- 6.8.2　能力目标 …… 154
- 6.8.3　任务驱动 …… 154
- 6.8.4　实践环节 …… 159

6.9　小结 …… 159

习题 6 …… 160

第 7 章　Java Servlet ……161

7.1　Servlet 类与 servlet 对象 ……161
- 7.1.1　核心知识 ……161
- 7.1.2　能力目标 ……161
- 7.1.3　任务驱动 ……162
- 7.1.4　实践环节 ……162

7.2　servlet 对象的创建与运行 ……163
- 7.2.1　核心知识 ……163
- 7.2.2　能力目标 ……166
- 7.2.3　任务驱动 ……166
- 7.2.4　实践环节 ……168

7.3　通过 JSP 页面访问 Servlet ……169
- 7.3.1　核心知识 ……169
- 7.3.2　能力目标 ……169
- 7.3.3　任务驱动 ……169
- 7.3.4　实践环节 ……171

7.4　doGet 和 doPost 方法 ……171
- 7.4.1　核心知识 ……171
- 7.4.2　能力目标 ……171
- 7.4.3　任务驱动 ……172
- 7.4.4　实践环节 ……174

7.5　重定向与转发 ……174
- 7.5.1　核心知识 ……174
- 7.5.2　能力目标 ……175
- 7.5.3　任务驱动 ……175
- 7.5.4　实践环节 ……177

7.6　在 Servlet 中使用 session ……178
- 7.6.1　核心知识 ……178
- 7.6.2　能力目标 ……178
- 7.6.3　任务驱动 ……178
- 7.6.4　实践环节 ……181

7.7　基于 Servlet 的 MVC 模式 ……181
- 7.7.1　核心知识 ……181
- 7.7.2　能力目标 ……181
- 7.7.3　任务驱动 ……181
- 7.7.4　实践环节 ……186

7.8　小结 ……186
习题 7 ……186

第 8 章　过滤器 ……188

8.1　Filter 类与 filter 对象 ……188
- 8.1.1　核心知识 ……188
- 8.1.2　能力目标 ……188

　　　　8.1.3　任务驱动 …………………………………………………………………… 189
　　　　8.1.4　实践环节 …………………………………………………………………… 190
　　8.2　filter 对象的部署与运行 ………………………………………………………………… 190
　　　　8.2.1　核心知识 …………………………………………………………………… 190
　　　　8.2.2　能力目标 …………………………………………………………………… 190
　　　　8.2.3　任务驱动 …………………………………………………………………… 190
　　　　8.2.4　实践环节 …………………………………………………………………… 193
　　8.3　过滤器的应用 …………………………………………………………………………… 193
　　　　8.3.1　核心知识 …………………………………………………………………… 193
　　　　8.3.2　能力目标 …………………………………………………………………… 194
　　　　8.3.3　任务驱动 …………………………………………………………………… 194
　　　　8.3.4　实践环节 …………………………………………………………………… 198
　　8.4　小结 ……………………………………………………………………………………… 198
　　习题 8 ………………………………………………………………………………………… 198

第 9 章　EL 与 JSTL ………………………………………………………………………… 199

　　9.1　表达式语言 EL …………………………………………………………………………… 199
　　　　9.1.1　核心知识 …………………………………………………………………… 199
　　　　9.1.2　能力目标 …………………………………………………………………… 203
　　　　9.1.3　任务驱动 …………………………………………………………………… 203
　　　　9.1.4　实践环节 …………………………………………………………………… 205
　　9.2　JSP 标准标签库 JSTL ……………………………………………………………………… 205
　　　　9.2.1　核心知识 …………………………………………………………………… 205
　　　　9.2.2　能力目标 …………………………………………………………………… 210
　　　　9.2.3　任务驱动 …………………………………………………………………… 211
　　　　9.2.4　实践环节 …………………………………………………………………… 212
　　9.3　小结 ……………………………………………………………………………………… 212
　　习题 9 ………………………………………………………………………………………… 213

第 10 章　文件的上传与下载 ………………………………………………………………… 214

　　10.1　基于 Servlet 3.0 的文件上传 …………………………………………………………… 214
　　　　10.1.1　核心知识 ………………………………………………………………… 214
　　　　10.1.2　能力目标 ………………………………………………………………… 217
　　　　10.1.3　任务驱动 ………………………………………………………………… 217
　　　　10.1.4　实践环节 ………………………………………………………………… 222
　　10.2　文件的下载 …………………………………………………………………………… 222
　　　　10.2.1　核心知识 ………………………………………………………………… 222
　　　　10.2.2　能力目标 ………………………………………………………………… 222
　　　　10.2.3　任务驱动 ………………………………………………………………… 223
　　　　10.2.4　实践环节 ………………………………………………………………… 229
　　10.3　小结 …………………………………………………………………………………… 229
　　习题 10 ……………………………………………………………………………………… 229

第 11 章 地址簿管理信息系统 ……………………………………………………… 230

11.1 系统设计 ………………………………………………………………………… 230
11.1.1 系统功能需求 ……………………………………………………………… 230
11.1.2 系统模块划分 ……………………………………………………………… 230

11.2 数据库设计 ……………………………………………………………………… 231
11.2.1 数据库概念结构设计 ……………………………………………………… 231
11.2.2 数据库逻辑结构设计 ……………………………………………………… 232
11.2.3 创建数据表 ………………………………………………………………… 232

11.3 系统管理 ………………………………………………………………………… 233
11.3.1 导入相关的 jar 包 ………………………………………………………… 233
11.3.2 JSP 页面管理 ……………………………………………………………… 233
11.3.3 组件与 Servlet 管理 ……………………………………………………… 235

11.4 组件设计 ………………………………………………………………………… 236
11.4.1 过滤器 ……………………………………………………………………… 236
11.4.2 数据库操作 ………………………………………………………………… 238
11.4.3 实体模型 …………………………………………………………………… 239
11.4.4 业务模型 …………………………………………………………………… 239

11.5 系统实现 ………………………………………………………………………… 245
11.5.1 用户注册 …………………………………………………………………… 245
11.5.2 用户登录 …………………………………………………………………… 247
11.5.3 添加朋友信息 ……………………………………………………………… 249
11.5.4 查询朋友信息 ……………………………………………………………… 253
11.5.5 查看详情 …………………………………………………………………… 255
11.5.6 修改朋友信息 ……………………………………………………………… 257
11.5.7 删除朋友信息 ……………………………………………………………… 262
11.5.8 修改密码 …………………………………………………………………… 263
11.5.9 退出系统 …………………………………………………………………… 265

参考文献 ………………………………………………………………………………… 266

第1章 Web 前端基础

主要内容

(1) HTML。
(2) CSS。
(3) JavaScript。

HTML 的英文全称是 Hyper Text Markup Language，即超文本标记语言，它是 Internet 上用于编写网页的主要标记语言。

CSS 是英文 Cascading Style Sheet 的缩写，又称为"层叠样式表"，简称为样式表。它是 W3C 定义的标准，一种用来为结构化文档（如 HTML 文档）添加样式（字体、间距和背景等）的计算机语言。CSS 是对 HTML 处理样式的补充，能将内容和样式处理相分离，大大降低了工作量。

JavaScript 是一种描述性的脚本语言（Script Language），它由客户端浏览器解释执行，执行期间无须 Web 服务器，减轻了 Web 服务器的负担。JavaScript 可以向 HTML 页面添加交互行为、读写元素、验证表单以及事件处理。

1.1 HTML

1.1.1 核心知识

1. HTML 文件的基本结构

一个完整的 HTML 文件由各种元素与标记组成，包括标题、段落、表格、文本和超链接等。下面是一个 HTML 文件的基本结构。

```
<html>
    <head>
        ...
    </head>
    <body>
        ...
    </body>
</html>
```

从上面的代码段可以看出，HTML 文件的基本结构分为 3 部分，其中各部分含义如下：

<html>...</html>：表示 HTML 文件开始和结束的位置，里面包括 head 和 body 等标记。HTML 文件中所有的内容都应该在这两个标记之间。

<head>...</head>：HTML 文件的头部标记，习惯将这两个标记之间的内容统称为 HTML 的头部。

<body>...</body>：用来指明文档的主体区域，网页所要显示的内容都要放置在这个标记内。习惯将这两个标记之间的内容统称为 HTML 的主体。

2. 编写 HTML 页面

编写 HTML 页面有两种常用方法：一种是利用操作系统自带的记事本编写；另一种是利用可视化网页制作软件（如 Dreamweaver）编写。本书从第 2 章开始使用集成开发环境（IDE）Eclipse 编写 Web 程序。

HTML、CSS 与 JavaScript 并不需要特殊的开发环境，它们都是由客户端的浏览器执行。HTML 文件的扩展名为.html 或.htm，CSS 文件的扩展名为.css，JavaScript 文件的扩展名为.js。

3. 常用 HTML 标记

常用 HTML 标记简单划分为以下 4 种格式。

(1) <标记名称>：单一型，无设置值。例如：
。
(2) <标记名称 属性="属性值">：单一型，有设置值。例如：<hr color="red">。
(3) <标记名称>...</标记名称>：对称型，无设置值。例如：<title>...</title>。
(4) <标记名称 属性="属性值">...</标记名称>：对称型，有设置值。例如：<body bgcolor="red">...</body>。

下面介绍常用的 HTML 标记。

1) 标题

HTML 将和文本相关的标题分成 6 个级别，1～6 级的标题语法格式如下：

```
<h1>…</h1>
<h2>…</h2>
<h3>…</h3>
<h4>…</h4>
<h5>…</h5>
<h6>…</h6>
```

h1 到 h6，作为标题标记，并且依据重要性递减，字号从 h1 到 h6 由大变小。为了更好地理解，请看下面的代码段：

```
<h1>学习标题标记</h1>
    <h2>第 1 章 Web 前端基础</h2>
        <h3>1.1HTML</h3>
            <h4>1.1.3 常用 HTML 标记</h4>
```

h1 一级标题代表重中之重，一般运用于网站标题或者头条新闻上。h2 二级标题主要出现在页面的主体内容的文章标题和栏目标题上。h3 三级标题一般出现在页面的边侧栏上。页面层级关系不能太深，所以 h4、h5 和 h6 一般出现得较少。

2) 段落

在 HTML 网页中,使用 p 标记实现一个新段落,语法格式如下:

<p>段落的内容</p>

p 标记中有一个属性 align 能够设置段落中文字的对齐方式,对齐方式分为左对齐、居中和两端对齐,语法格式如下:

<p align = "对齐方式"></p>

其中,align 取值为 left 时,文字显示左对齐;align 取值为 right 时,文字显示右对齐;align 取值为 center 时,文字显示居中对齐。

【例 1.1】 有 3 段文字,对齐方式依次为左、中、右。代码段如下:

```
<p align = "left">居左文字</p>
<p align = "center">居中文字</p>
<p align = "right">居右文字</p>
```

3) 滚动

在 HTML 页面中,可以使用 marquee 标记让文字滚动,该标记有滚动方向(direction)、滚动方式(behavior)、滚动次数(loop)、滚动速度(scrollamount)、滚动延迟(scrolldelay)、背景颜色(bgcolor)、宽度和高度等常用属性。语法格式如下:

< marquee direction = "滚动方向" behavior = "滚动方式">滚动的文字</marquee >

其中,direction 的值有 up、down、left 和 right,分别表示向上、向下、向左和向右滚动,向左滚动是默认情况;behavior 的值有 scroll、slide 和 alternate,分别表示循环滚动、只滚动一次和来回交替滚动;loop 的值为整数;scrollamount 的值为文字每次移动的长度,以像素为单位;scrolldelay 的单位是毫秒。

【例 1.2】 编写一个网页,网页中有一段滚动的文字,文字滚动方向为默认方向,文字滚动的背景色为蓝色,文字滚动方式为来回交替滚动。网页运行效果如图 1.1 所示。

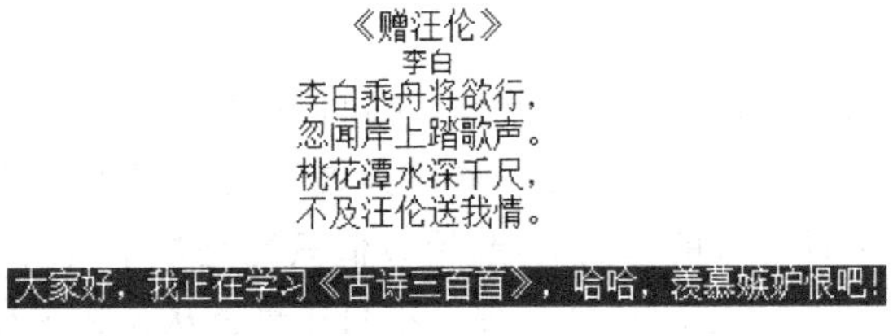

《赠汪伦》
李白
李白乘舟将欲行,
忽闻岸上踏歌声。
桃花潭水深千尺,
不及汪伦送我情。

大家好,我正在学习《古诗三百首》,哈哈,羡慕嫉妒恨吧!

图 1.1 example1_2.html 运行效果

例 1.2 页面文件 example1_2.html 的代码如下:

```
<!DOCTYPE html PUBLIC " - //W3C//DTD HTML 4.01 Strict//EN"
"http://www.w3.org/TR/html4/strict.dtd">
<html>
    <head>
    <title> example1_2.html </title>
    </head>
```

```
    <body>
        <p align = "center">«赠汪伦»<br><font size = "2">李白</font><br>
            李白乘舟将欲行,<br>
            忽闻岸上踏歌声.<br>
            桃花潭水深千尺,<br>
            不及汪伦送我情.<br>
        </p>
        <marquee bgcolor = "blue" behavior = "alternate">
            <font color = "white">大家好,我正在学习«古诗三百首»,哈哈,羡慕嫉妒恨吧!</font>
        </marquee>
    </body>
</html>
```

4)列表

(1)无序列表标记 ul。ul 标记用于设置无序列表,在每个列表项目文字之前,以项目符号作为每条列表项的前缀,各个列表没有级别之分。无序列表语法格式如下:

```
<ul>
    <li>列表项</li>
    <li>列表项</li>
    …
</ul>
```

无序列表的项目符号默认情况下是"●",而通过 ul 标记的 type 属性可以改变无序列表的项目符号,避免项目符号的单调。type 可取值 disc、circle 和 square,分别代表"●""○"和"■"。

(2)有序列表标记 ol。有序列表中的项目采用数字或英文字母开头,通常各项目之间是有先后顺序的。有序列表语法格式如下:

```
<ol>
    <li>列表项</li>
    <li>列表项</li>
    …
</ol>
```

有序列表同无序列表一样,也有项目类型,也可以通过 type 属性设置自己的项目类型。默认情况下,有序列表的项目序号是数字。有序列表 type 属性的取值如表 1.1 所示。也可以通过 start 属性改变项目序号的起始值,起始数值只能是数字,但同样对字母或罗马数字起作用。例如,项目类型为 a,起始值为 5,那么项目序号就从英文字母 e 开始编号。

表 1.1 有序列表 type 属性的取值

属性值	项目序号	属性值	项目序号
1	1、2、3、4…	i	i、ii、iii、iv…
a	a、b、c、d…	I	Ⅰ、Ⅱ、Ⅲ、Ⅳ…
A	A、B、C、D…		

【例 1.3】 编写一个网页(运行效果如图 1.2 所示),在网页中分别定义一个无序列表和一个有序列表,无序列表项目符号为"○",有序列表项目序号为"a、b、c、d…"。

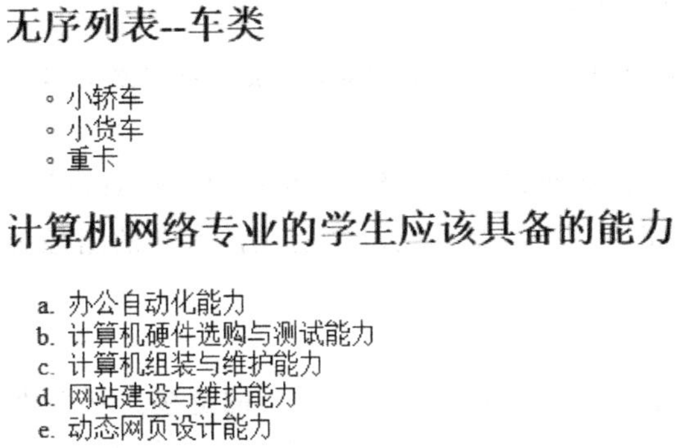

无序列表--车类
  。小轿车
  。小货车
  。重卡

计算机网络专业的学生应该具备的能力
  a. 办公自动化能力
  b. 计算机硬件选购与测试能力
  c. 计算机组装与维护能力
  d. 网站建设与维护能力
  e. 动态网页设计能力

图 1.2　example1_3.html 运行效果

例 1.3 页面文件 example1_3.html 的代码如下：

```html
<!DOCTYPE html PUBLIC "-//W3C//DTD HTML 4.01 Strict//EN"
"http://www.w3.org/TR/html4/strict.dtd">
<html>
    <head>
    <title>example1_3.html</title>
    </head>
    <body>
        <h2>无序列表 -- 车类</h2>
        <ul type="circle">
            <li>小轿车</li>
            <li>小货车</li>
            <li>重卡</li>
        </ul>
        <h2>计算机网络专业的学生应该具备的能力</h2>
        <ol type="a">
            <li>办公自动化能力</li>
            <li>计算机硬件选购与测试能力</li>
            <li>计算机组装与维护能力</li>
            <li>网站建设与维护能力</li>
            <li>动态网页设计能力</li>
        </ol>
    </body>
</html>
```

5）图像与多媒体

HTML 图像是通过 img 标记插入的。img 标记有很多属性,其中 src 属性是必需的,它指定要插入图像文件的位置与名称,语法格式如下：

`<img src="图像文件的路径及名称">`

在网页中可以使用 bgsound 标记给网页添加背景音乐,语法格式如下：

`<bgsound src="音乐文件的路径及名称" loop="播放次数">`

在网页中可以使用 embed 标记将多媒体文件（如 Flash 动画、MP3 音乐、ASF 视频等）

添加到网页中,语法格式如下:

<embed src = "多媒体文件的路径及名称" width = "播放器的宽度" height = "播放器的高度"></embed>

图像与多媒体文件的路径可以是相对路径,也可以是绝对路径。绝对路径是完全路径,是文件在硬盘上的真正路径。相对路径是以当前文件所在的路径为基准,进行相对文件的查找。

6) 超链接

超链接的作用是建立从一个位置到另一个位置的链接。利用超链接不仅可以进行网页间的相互访问,还可以使网页链接到其他相关的多媒体文件上。

超链接标记 a 是一个非常重要的标记,它可以成对出现在文档的任何位置,语法格式如下:

<a href = "链接路径" target = "目标窗口的打开方式">链接内容</a>

其中,"链接内容"可以是文字内容,也可以是一张图片。target 属性值可以为_self、_blank、_top 以及_parent,_self 是 target 的默认值。_blank 表示目标页面会在一个新的空白窗口中打开。_top 表示目标页面会在顶层框架中打开。_parent 表示目标页面会在当前框架的上一层打开。

【例 1.4】 假设有 3 个文件,分别为 index.html、addGoods.html 和 updateGoods.html。其中 index.html 是起始页面,addGoods.html 和 updateGoods.html 在 goods 文件夹下,goods 文件夹和 index.html 在同一目录。在 index.html 中可以链接到后面两个页面上。

index.html 的代码如下:

```
<!DOCTYPE html PUBLIC " - //W3C//DTD HTML 4.01 Strict//EN"
"http://www.w3.org/TR/html4/strict.dtd">
<html>
    <head>
    <title>电子商务后台首页</title>
    </head>
    <body>
        <ul>
            <li><a href = "index.html">首页</a></li>
            <li><a href = "goods/addGoods.html">添加商品</a></li>
            <li><a href = "goods/updateGoods.html">修改商品</a></li>
        </ul>
        <p>首页</p>
    </body>
</html>
```

在例 1.4 中,由于 addGoods.html、updateGoods.html 和 index.html 的相对路径为 goods/,在 addGoods.html 中需要跳转到 index.html 和 updateGoods.html 两个页面上。

addGoods.html 的代码如下:

```
<!DOCTYPE html PUBLIC " - //W3C//DTD HTML 4.01 Strict//EN"
```

```
"http://www.w3.org/TR/html4/strict.dtd">
<html>
    <head>
    <title>添加商品</title>
    </head>
    <body>
        <ul>
            <li><a href="../index.html">首页</a></li>
            <li><a href="addGoods.html">添加商品</a></li>
            <li><a href="updateGoods.html">修改商品</a></li>
        </ul>
        <p>添加商品页面</p>
    </body>
</html>
```

由于起始页面 index.html 位于当前页面(addGoods.html)的上一级,所以从当前页面到起始页面的链接写为../index.html。而 updateGoods.html 和 addGoods.html 在同一级别目录下,可以省略路径名。

7) 表格

一个表格由行、列和单元格构成,可以有多行,每行可以有多个单元格。创建表格要以<table>标记开始,以</table>标记结束,语法格式如下:

```
<table>
    <tr>
        <td>单元格中的内容</td>
        <td>单元格中的内容</td>
         …
    </tr>
    <tr>
        <td>单元格中的内容</td>
        <td>单元格中的内容</td>
         …
    </tr>
     …
</table>
```

在一个表格中包含几组<tr>和</tr>标记,就表示该表格有几行。在一行中包含几组<td>和</td>标记,就表示该行中有几个单元格。

在制作表格时,可能需要某个单元格占据多列的位置,这时就要使用单元格的 colspan 属性设置单元格所跨的列数,语法格式如下:

<td colspan="跨的列数值">

如下表第一行的单元格的水平跨度为 5。

同样地，当需要某个单元格占据多行的位置时，就要使用 rowspan 属性设置该单元格所跨的行数。

```
<td rowspan = "跨的行数值">
```

如下表第一个单元格的垂直跨度为 2。

<table>
<tr><td rowspan="2"></td><td></td><td></td><td></td><td></td></tr>
<tr><td></td><td></td><td></td><td></td></tr>
</table>

【例 1.5】 编写网页 example1_5. html，在网页中有一个表格，表格标题为“个人简历”，边框宽度为 1，边框颜色为 green。网页运行效果如图 1.3 所示。

<table>
<caption>个人简历</caption>
<tr><th rowspan="2">基本资料</th><th>姓名</th><td>顾小白</td><th>性别</th><td>女</td></tr>
<tr><th>政治面貌</th><td>群众</td><th>出生日期</th><td>1988-12-09</td></tr>
<tr><th colspan="5">业余爱好</th></tr>
<tr><td>读书</td><td>各种球类（包括中国足球）</td><td>看电影</td><td>爬山</td><td>逛街</td></tr>
</table>

图 1.3　example1_5. html 运行效果

例 1.5 页面文件 example1_5. html 的代码如下：

```
<!DOCTYPE html PUBLIC " - //W3C//DTD HTML 4.01 Strict//EN"
"http://www.w3.org/TR/html4/strict.dtd">
<html>
    <head>
    <title>example1_5.html</title>
    </head>
    <body>
        <table border = "1" bordercolor = "green">
            <caption>个人简历</caption>
            <tr>
                <th rowspan = "2" align = "left">基本资料</th>
                <th align = "left">姓名</th>
                <td>顾小白</td>
                <th align = "left">性别</th>
                <td>女</td>
            </tr>
            <tr>
                <th align = "left">政治面貌</th>
                <td>群众</td>
                <th align = "left">出生日期</th>
                <td>1988 - 12 - 09</td>
            </tr>
            <tr>
                <th colspan = "5">业余爱好</th>
            </tr>
            <tr>
                <td>读书</td>
                <td>各种球类(包括中国足球)</td>
                <td>看电影</td>
```

```
            <td>爬山</td>
            <td>逛街</td>
        </tr>
    </table>
</body>
</html>
```

8) 表单

表单是将用户输入的信息封装提交给服务器,实现与用户的交互。表单是用 form 标记定义的,它类似于一个容器,表单对象必须在表单中才有效,如 input。定义表单的语法格式如下:

```
<form action="表单的处理程序">
    …
    input 元素
    …
</form>
```

form 标记有很多属性,如表 1.2 所示。

表 1.2 form 标记的属性及含义

属性名	含 义 描 述
action	action 属性值是表单提交的地址,即表单收集到信息后传递到的程序地址,如某页面
name	name 属性给表单命一个名称。表单名称可以控制表单与后台程序之间的关系
method	method 属性用于指定使用哪种提交方法将表单数据提交到服务器。默认情况下,提交方法为 get。get 方法将表单内容附在 URL 地址后面,因此有长度限制(最大 8192 个字符),而且不安全。post 方法将用户在表单中输入的数据包含在表单主体中,一起提交给服务器,该方法没有信息长度的限制,也比较安全
enctype	enctype 属性用于指定表单信息提交的编码方式,这个编码方式通常情况下采用默认的(application/x-www-form-urlencoded)即可,但上传文件时必须选择 mime 编码(multipart/form-data)

input 标记是最常用的表单标记,该标记允许用户在表单中(文本框、单选框、复选框等)输入信息,输入类型是由类型属性(type)定义的。常用的输入类型如下。

(1) 文本框与密码框。input 标记的 type 属性值为 text,代表的是单行文本框,在其中可以输入任何类型的文本、数字或字母,输入的内容以单行显示。input 标记的 type 属性值为 password,代表的是密码框,在其中输入的字符都是以"﹡"或圆点"●"显示。如下面原始代码:

```
<form>
    姓名:<input type="text" value="" name="userName"/>
    <br>
    密码:<input type="password" value="123456" name="pwd"/>
</form>
```

图 1.4 文本框与密码框

上述原始代码呈现的结果如图 1.4 所示。

(2) 单选按钮与复选框。单选按钮用来让用户进行单一选择,如让用户选择性别。单

选按钮在页面中以圆框("○")显示,语法格式如下:

```
< input type = "radio" value = "单选按钮的值" name = "单选按钮的名称" checked/>
```

其中,name 代表单选按钮的名称,一组单选按钮的名称都相同,action 处理程序通过 name 获取被选中的单选按钮的 value 值。checked 表示该单选按钮被选中,而在一组单选按钮中只有一个单选按钮设置为 checked。如下面的原始代码:

```
< form >
    < input type = "radio" value = "male" name = "sex" checked/>男
    < input type = "radio" value = "female" name = "sex" />女
</form>
```

◉男 ○女

图1.5 单选按钮

上述原始代码呈现的结果如图 1.5 所示。

经常看到这样的问题,请选择您喜欢的歌手:□张三□李四□王五,这样的网页就是使用复选框实现的。复选框与单选按钮不同的是复选框能够实现选项的多选,以一个方框("□")表示,语法格式如下:

```
< input type = "checkbox" value = "复选框的值" name = "复选框的名称" checked/>
```

其中,当用户勾选复选框后,value 属性的值传递给处理程序。name 代表的是复选框的名称,一组复选框的名称都相同,处理程序通过 name 获取被选中的复选框的 value 值(以数组的形式返回,数组元素为被选中的复选框的 value 值)。checked 表示该复选框被选中,一组复选框中可以同时有多个被选中。如下面的原始代码:

```
< form >
    < input type = "checkbox" value = "zhangsan" name = "lover" checked/>张三
    < input type = "checkbox" value = "lisi" name = "lover" checked/>李四
    < input type = "checkbox" value = "wangwu" name = "lover"/>王五
</form>
```

☑张三 ☑李四 □王五

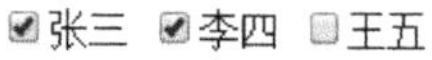

图1.6 复选框

上述原始代码呈现的结果如图 1.6 所示。

(3)按钮。在网页的表单中,按钮起到至关重要的作用。没有按钮,网页很难和用户进行交互。单击按钮可以激发提交表单的动作(提交按钮);也可以将表单恢复到初始的状态(重置按钮);还可以根据程序的要求,发挥其他的作用(普通按钮)。

普通按钮主要是配合脚本语言(JavaScript)进行表单的处理,语法格式如下:

```
< input type = "button" value = "按钮的值" name = "按钮的名称"/>
```

其中,value 的取值就是显示在按钮上的文字,在普通按钮中可以添加 onclick、onfocus 等 JavaScript 事件实现特定的功能。

当用户在表单中输入信息后,想清除输入的信息,将表单恢复成初始状态时,需要使用重置按钮,语法格式如下:

```
< input type = "reset" value = "按钮的值" name = "按钮的名称"/>
```

用户单击提交按钮时,可以实现表单内容的提交,语法格式如下:

```
< input type = "submit" value = "按钮的值" name = "按钮的名称"/>
```

如下面的原始代码：

```
< form >
    姓名: < input type = "text" name = "userName"/><br>
    < input type = "submit" value = "提交"/>
    < input type = "reset" value = "重置"/>
    < input type = "button" value = "关闭" onclick = "window.close()"/>
</form>
```

上述原始代码呈现的结果如图1.7所示。

图1.7 按钮

（4）文件域。经常需要实现上传照片或文件的功能，这就用到文件域。文件域是由一个文本框和一个"浏览"按钮组成的。用户上传文件时，可以直接在文本框中输入要上传文件的路径，也可以单击"浏览"按钮选择文件，语法格式如下：

```
< input type = "file" name = "文件域的名称"/>
```

使用文件域上传文件时，一定不要忘记设置form表单信息提交的编码方式为enctype="multipart/form-data"。如下面的原始代码：

图1.8 文件域

```
< form action = "" enctype = "multipart/form-data">
    你的靓照: < input type = "file" name = "fileName"/>
</form>
```

上述原始代码呈现的结果如图1.8所示。

（5）下拉列表。下拉列表语法格式如下：

```
< select name = "下拉列表的名称" size = "显示的项数" multiple >
    < option value = "选项值1" selected>选项1显示内容
    < option value = "选项值2" >选项2显示内容
    ...
</select >
```

其中，选项值是提交给服务器的值，而选项显示内容才是真正在页面中要显示的。selected表示此选项在默认状态下是选中的，size用来设定列表在页面中最多显示的项数，当超出这个值时就会出现滚动条。multiple表示列表可以进行多项选择。如下面的原始代码：

```
< select name = "cities" size = "2" multiple >
    < option value = "beijing" selected>北京
    < option value = "shanghai" selected>上海
    < option value = "guangzhou" >广州
    < option value = "shenzhen" >深圳
</select >
```

图1.9 列表

上述原始代码呈现的结果如图1.9所示。

（6）文本区。文本区用来输入多行文本。文本区和其他表单控件不一样，它使用的是textarea标记而不是input标记，语法格式如下：

```
<textarea name="文本区的名称" cols="列数" rows="行数"></textarea>
```

其中,cols 用于设定文本区的列数,也就是其宽度;rows 用于设定文本区的行数,也就是高度值,当文本区的内容超出这一范围时就会出现滚动条。

1.1.2 能力目标

通过本节的学习,掌握 HTML 文件的基本结构、基本标签,能够编写简单的 HTML 页面。

1.1.3 任务驱动

1. 任务的主要内容

编写一个 HTML 文件 task1_1.html,在该文件中嵌套使用无序列表和有序列表。HTML 文件运行效果如图 1.10 所示。

2. 任务的代码模板

task1_1.html 的代码模板如下:

```
<!DOCTYPE html PUBLIC "-//W3C//DTD HTML 4.01 Strict//EN"
"http://www.w3.org/TR/html4/strict.dtd">
<html>
    <head>
    <meta http-equiv="Content-Type" content="text/html; charset=UTF-8">
    <title>序列表的嵌套使用</title>
    </head>
    <body>
        饮品:
        【代码1】
            <li>咖啡</li>
            <li>茶
                【代码2】
                    <li>红茶</li>
                    <li>绿茶</li>
                【代码3】
            </li>
            <li>牛奶</li>
        【代码4】
    </body>
</html>
```

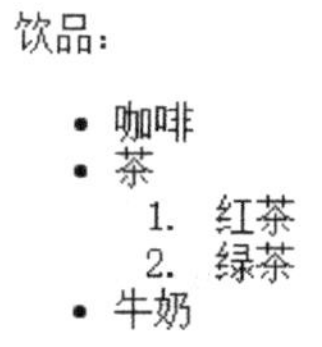

饮品:
- 咖啡
- 茶
    1. 红茶
    2. 绿茶
- 牛奶

图 1.10 嵌套使用无序列表和有序列表

3. 任务小结或知识扩展

不管是无序列表还是有序列表,它们都可以任意地嵌套使用。例如,任务中无序列表嵌套使用有序列表。

4. 代码模板的参考答案

【代码 1】:＜ul＞
【代码 2】:＜ol＞
【代码 3】:＜/ol＞
【代码 4】:＜/ul＞

1.1.4 实践环节

编写网页 practice1_1.html,具体要求如下。

(1) 网页中有一个 form 表单,表单处理程序为本页面程序,表单提交方式为 post,表单提交编码方式为 multipart/form-data。

(2) 页面运行效果如图 1.11 所示。

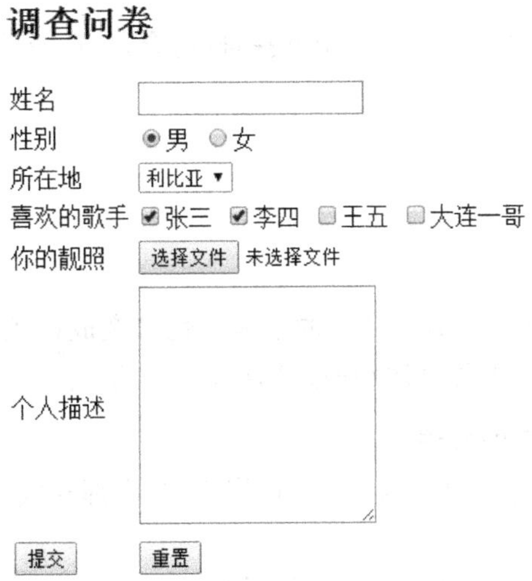

图 1.11　practice1_1.html 页面运行效果

1.2　CSS

不需要使用复杂的工具来创建 CSS 文件,可以使用文本编辑器或者 Web 开发工具来创建。无论采用哪种方式,都是要创建一个以.css 为扩展名的文件。

1.2.1 核心知识

1. CSS 基本语法

CSS 的语法由 3 部分构成：选择符(selector)、属性(property)和属性值(value)。语法格式如下：

```
选择符{
    属性:值
}
```

选择符用来指定针对哪个 HTML 标签应用样式表,任何一个 HTML 标签都可以是一个 CSS 的选择符。例如：

```
body {
    color: blue
}
```

其中,body 就是选择符,color 就是属性,blue 就是属性值。该规则表示在网页 body 标签里的内容为蓝色。为选择符指定多个样式,需要在属性之间用分号加以分隔。下面的选择符 p 就包含两个样式,一个是对齐方式为居中,一个是字体颜色为红色。

```
p {
    text-align: center;
    color: red
}
```

可以将相同的属性和属性值赋予多个选择符(组合)。选择符之间用逗号分隔。例如：

```
h1,h2,h3,h4,h5,h6
{
    /*字体颜色为蓝色*/
    color: blue
}
```

该规则是将所有正文标题(h1 到 h6)的字体颜色都变成蓝色。"/*"和"*/"之间的内容为 CSS 的注释,但要注意不要将注释嵌入选择符语句里面。

2. 在网页中添加 CSS 的方法

CSS 在网页中按其位置可以分为 3 种：内嵌样式、内部样式和外部样式。

1) 内嵌样式

内嵌样式是将样式代码写在标记里面的,使用 style 作为属性,样式语句作为属性值。内嵌样式只对所在标记有效。例如：

```
<p style="font-size:20pt; color:red">
    这个 style 定义<p></p>里面的文字是 20pt 字号,字体颜色是红色
</p>
```

2) 内部样式

内部样式是使用 style 标记将样式代码写在 HTML 的<head></head>里面的。内部样式只对所在网页有效。例如：

```
<html>
    <head>
        <style type="text/css">
            h1 {
                border-width:1;
                text-align:center;
                color:red
            }
        </style>
```

```
        </head>
        <body>
            <h1>这个标题使用了 style</h1>
        </body>
</html>
```

3) 外部样式

(1) 链接样式表。将样式代码写在一个以.css 为后缀的 CSS 文件里,然后在每个需要用到这些样式的网页里引用这个 CSS 文件。通过 HTML 的 link 元素将外部的样式文件链接到网页里。例如:

```
<!DOCTYPE html PUBLIC "-//W3C//DTD HTML 4.01 Strict//EN"
"http://www.w3.org/TR/html4/strict.dtd">
<html>
    <head>
    <meta http-equiv = "Content-Type" content = "text/html; charset = UTF-8">
    <title>Insert title here</title>
    <link rel = "stylesheet" href = "mystyle.css" type = "text/css" />
    </head>
    <body>
        <h1>标题</h1>
        <p>段落内容</p>
    </body>
</html>
```

其中,rel 和 type 属性表明这是一个样式文件,href 属性指定了外部样式文件的相对地址。外部的样式文件不能含有任何像 head 或 style 这样的 HTML 标记,样式表文件仅仅由样式规则或声明组成。mystyle.css 文件内容如下:

```
p{
    background: yellow;
}
```

(2) 导入样式表。在 style 标记内,使用@import 导入外部样式文件。例如:

```
<!DOCTYPE html PUBLIC "-//W3C//DTD HTML 4.01 Strict//EN"
"http://www.w3.org/TR/html4/strict.dtd">
<html>
    <head>
    <meta http-equiv = "Content-Type" content = "text/html; charset = UTF-8">
    <title>Insert title here</title>
        <style type = "text/css">
            <!--
            @import url("mystyle.css");
            h1{color:red}
            -->
        </style>
    </head>
    <body>
        <h1>标题</h1>
        <p>段落内容</p>
```

```
</body>
</html>
```

使用外部样式,相对于内嵌样式和内部样式,有以下优点。

① 样式代码可以复用。一个外部 CSS 文件,可以被很多网页共用。

② 便于修改。如果要修改样式,只需要修改 CSS 文件,而不需要修改每个网页。

③ 提高网页显示的速度。如果样式写在网页里,会降低网页显示的速度,如果网页引用一个 CSS 文件,这个 CSS 文件已经在缓存区(其他网页早已经引用过它),网页显示的速度就比较快。

因此,在实际开发中一般使用外部样式,不推荐使用内嵌样式和内部样式。

3. 选择符的分类

1) 普通选择符

任意的 HTML 标记都可以作为选择符,这样的选择符称为普通选择符。其样式仅作用在选择符指定的 HTML 标记上。例如:

```
p {
    color:red
}
```

2) 类选择符

HTML 标记名加上类名,中间用"."号分开,类名供该 HTML 标记的 class 属性使用。如果希望 p 标记有两种样式,一种是居中对齐,另一种是居右对齐,那么可以写成如下的样式:

```
p.right {
    text-align:right
}
p.center {
    text-align:center
}
```

其中,right 和 center 就是两个 class。在网页中可以引用这两个 class 设置段落的对齐方式。示例代码如下:

```
<p class = "center">这一段内容是居中显示</p>
<p class = "right">这一段内容是居右显示</p>
```

如果省略 HTML 标记名只写".类名"表示这个类名适用于所有的 HTML 标记的 class 属性,这种选择符称为通用类选择符。

3) id 选择符

HTML 标记名加上 id 名,中间用"#"号分开,id 名供该 HTML 标记的 id 属性使用。例如:

```
p#svp {
    font-size:12pt
}
```

其中,svp 是一个 id 选择符的名字,在网页中可以引用这个 id 选择符设置 p 标记的样式。示例代码如下:

```
<p id = "svp">这一段话的字体大小为12pt.</p>
```

如果省略 HTML 标记名只写"♯id 名"表示这个 id 名适用于所有的 HTML 标记的 id 属性,这种选择符称为通用 id 选择符。

4. 伪类及伪对象

伪类及伪对象由 CSS 自动支持,属于 CSS 的一种扩展类型。名称不能被用户自定义,使用时只能按照标准格式进行应用。

1) 超链接伪类

超链接伪类共有 4 个,它们是 a:link、a:visited、a:hover 和 a:active。a:link 表示未被访问的链接;a:visited 表示已被访问过的链接;a:hover 表示鼠标悬浮在上的链接;a:active 表示鼠标点中激活的链接。由于优先级的关系,在写超链接 a 标记的 CSS 时,一定要按照 a:link、a:visited、a:hover、a:active 的顺序书写。例如:

```
a:link {color: red}          /* 未被访问的链接 红色 */
a:visited {color: green}     /* 已被访问过的链接 绿色 */
a:hover {color: yellow}      /* 鼠标悬浮在上的链接 黄色 */
a:active {color: blue}       /* 鼠标点中激活的链接 蓝色 */
```

2) 常用伪对象

:first-letter 设置对象内的第一个字符的样式表属性,如设置段落 p 标记的第一个字符的样式代码如下:

```
p:first-letter {
    color: red;
    font-size: 16px
}
```

:first-line 设置对象内的第一行的样式表属性,如设置 body 对象里第一行的样式代码如下:

```
body:first-line {
    color: red;
    font-size: 16px
}
```

5. 常见的 DIV+CSS 布局类型

DIV+CSS 布局是当前网页布局中最流行的类型之一。

1) DIV

DIV 是一个放置内容的容器,用于大面积、大区域的块状排版,样式需要编写 CSS 来实现。示例代码如下:

```
<!DOCTYPE html PUBLIC "-//W3C//DTD HTML 4.01 Strict//EN"
"http://www.w3.org/TR/html4/strict.dtd">
<html>
    <head>
        <meta http-equiv = "Content-Type" content = "text/html; charset = UTF-8">
        <title>div_css1</title>
```

```
<style type="text/css">
.mainBox {
    border: 1px dashed #0099CC;
    margin: 3px;
    padding: 0px;
    float: left;
    height: 300px;
    width: 192px;
}
.mainBox h3 {
    float: left;
    height: 20px;
    width: 179px;
    color: #FFFFFF;
    padding: 6px 3px 3px 10px;
    background-color: #0099CC;
    font-size: 16px;
}
.mainBox p {
    line-height: 1.5em;
    text-indent: 2em;
    margin: 35px 5px 5px 5px;
}
</style>
</head>
<body>
    <div class="mainBox">
        <h3>前言</h3>
        <p>正文内容</p>
    </div>
    <div class="mainBox">
        <h3>CSS 盒子模式</h3>
        <p>正文内容</p>
    </div>
    <div class="mainBox">
        <h3>转变思想</h3>
        <p>正文内容</p>
    </div>
</body>
</html>
```

上述代码运行效果如图 1.12 所示。

2）1 列固定

宽度的属性值是固定像素，无论怎样改变浏览器窗口大小，DIV 的宽度都不改变。示例代码如下：

```
<!DOCTYPE html PUBLIC "-//W3C//DTD HTML 4.01 Strict//EN"
"http://www.w3.org/TR/html4/strict.dtd">
<html>
    <head>
```

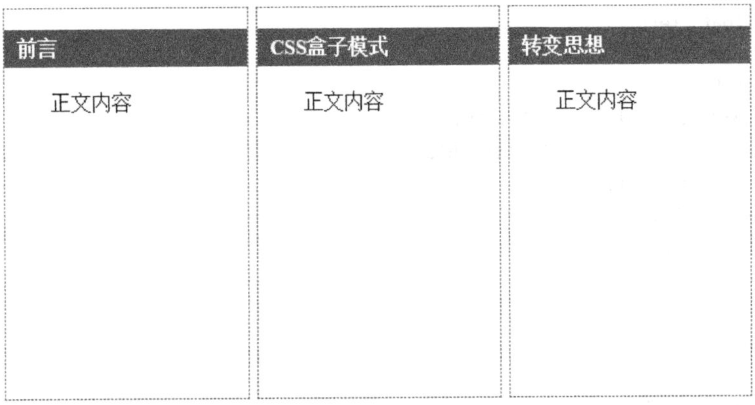

图 1.12　DIV 示例

```
    <meta http-equiv="Content-Type" content="text/html; charset=UTF-8">
    <title>div_css2</title>
    <style type="text/css">
    .oneFixed {
        border: 1px dashed #0099CC;
        background-color: cyan;
        height: 300px;
        width: 300px
    }
    </style>
    </head>
    <body>
        <div class="oneFixed">
            1列固定宽度
        </div>
    </body>
</html>
```

上述代码运行效果如图 1.13 所示。

图 1.13　1 列固定宽度

3）2 列固定宽度

2 列的布局需要用到两个 DIV，宽度的属性值是固定像素。示例代码如下：

```
<!DOCTYPE html PUBLIC "-//W3C//DTD HTML 4.01 Strict//EN"
"http://www.w3.org/TR/html4/strict.dtd">
<html>
    <head>
    <meta http-equiv="Content-Type" content="text/html; charset=UTF-8">
    <title>div_css3</title>
    <style type="text/css">
    #left {
        border: 1px dashed #0099CC;
        background-color: cyan;
        height: 200px;
```

```
            width: 300px;
            float: left
        }
        #right{
            border: 1px dashed #0099CC;
            background-color: LightSkyBlue;
            height: 200px;
            width: 300px;
            float: left
        }
        </style>
    </head>
    <body>
        <div id="left">
            左边
        </div>
        <div id="right">
            右边
        </div>
    </body>
</html>
```

上述代码运行效果如图 1.14 所示。

图 1.14 2 列固定宽度

4) 3 列浮动中间宽度自适应

3 列浮动中间宽度自适应就是要求左边 DIV 固定宽度且居左显示,右边 DIV 固定宽度且居右显示,中间 DIV 根据左右 DIV 的间距变化宽度自适应。示例代码如下:

```
<!DOCTYPE html PUBLIC "-//W3C//DTD HTML 4.01 Strict//EN"
"http://www.w3.org/TR/html4/strict.dtd">
<html>
    <head>
        <meta http-equiv="Content-Type" content="text/html; charset=UTF-8">
        <title>div_css4</title>
        <style type="text/css">
        body{
            margin: 0px
        }
        #left{
```

```
            border: 1px solid #0099CC;
            background-color: cyan;
            height: 300px;
            width: 100px;
            position: absolute;
            top: 0px;
            left: 0px
        }
        #center{
            border: 1px solid #0099CC;
            background-color: #7FFFAA;
            margin-left: 100px;
            margin-right: 100px;
            height: 300px
        }
        #right{
            border: 1px solid #0099CC;
            background-color: cyan;
            height: 300px;
            width: 100px;
            position: absolute;
            right: 0px;
            top: 0px
        }
    </style>
</head>
<body>
    <div id="left">
        左边
    </div>
    <div id="center">
        中间自适应
    </div>
    <div id="right">
        右边
    </div>
</body>
</html>
```

上述代码运行效果如图1.15所示。

5) 3行2列居中高度自适应

3行2列居中高度自适应就是要求整个网页内容居中，第一行DIV固定高度且居顶端显示，第3行DIV固定高度且居底端显示，中间DIV根据内容的变化高度自适应。示例代码如下：

```
<!DOCTYPE html PUBLIC "-//W3C//DTD HTML 4.01 Strict//EN"
"http://www.w3.org/TR/html4/strict.dtd">
```

图 1.15 3 列浮动中间宽度自适应

```
<html>
    <head>
    <meta http-equiv="Content-Type" content="text/html; charset=UTF-8">
    <title>div_css5</title>
    <style type="text/css">
    body {
        background: #999;
        text-align: center;
        color: #333;
        font-family: arial, verdana, sans-serif;
    }

    #header {
        width: 776px;
        margin-right: auto;
        margin-left: auto;
        padding: 0px;
        background: #EEE;
        height: 60px;
        text-align: left;
    }

    #contain {
        margin-right: auto;
        margin-left: auto;
        width: 776px;
    }

    #mainbg {
        width: 776px;
        padding: 0px;
        background: #60A179;
        float: left;
```

```css
}
#right {
    float: right;
    margin: 2px 0px 2px 0px;
    padding: 0px;
    width: 574px;
    background: #ccd2de;
    text-align: left;
}

#left {
    float: left;
    margin: 2px 2px 0px 0px;
    padding: 0px;
    background: #F2F3F7;
    width: 200px;
    text-align: left;
}

#footer {
    clear: both;
    width: 776px;
    margin-right: auto;
    margin-left: auto;
    padding: 0px;
    background: #EEE;
    height: 60px;
}

.text {
    margin: 0px;
    padding: 20px;
}
</style>
</head>
<body>
    <div id="header">header</div>
    <div id="contain">
        <div id="mainbg">
            <div id="right">
                <div class="text">
                    right
                    <p>1</p>
                    <p>1</p>
                    <p>1</p>
                    <p>1</p>
```

```
                    <p>1</p>
                </div>
            </div>
            <div id="left">
                <div class="text">left</div>
            </div>
        </div>
    </div>
    <div id="footer">footer</div>
</body>
</html>
```

上述代码运行效果如图 1.16 所示。

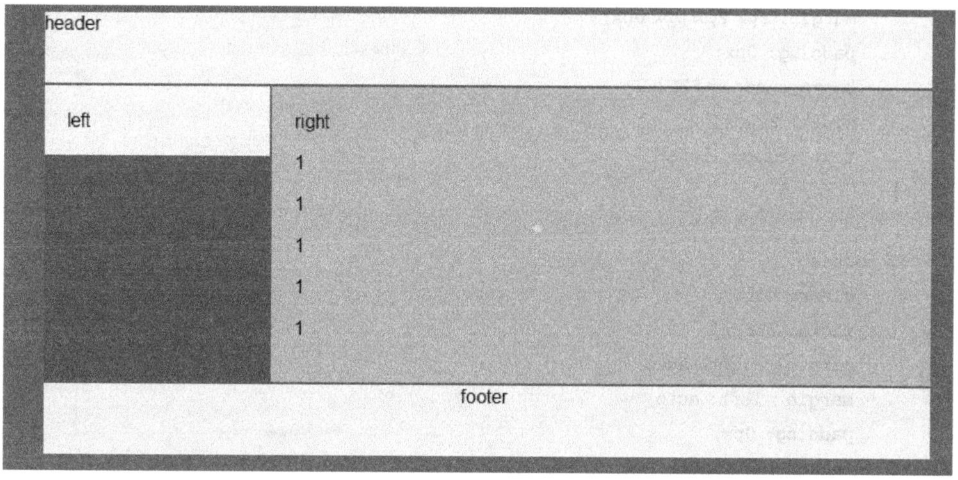

图 1.16　3 行 2 列居中高度自适应

1.2.2　能力目标

通过本节的学习，掌握 CSS 的基本语法和常用属性，能够编写简单的 CSS 样式文件。

1.2.3　任务驱动

1. 任务的主要内容

首先，编写一个 CSS 文件 myTask.css，在该 CSS 文件中分别定义一个类选择符和 id 选择符；然后，编写一个 HTML 文件 task1_2.html，在该 HTML 文件中使用 myTask.css 美化页面。task1_2.html 和 myTask.css 在同一个目录下。task1_2.html 的运行效果如图 1.17 所示。

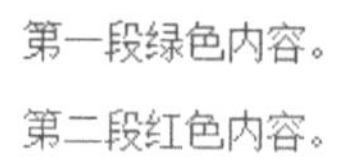

2. 任务的代码模板

myTask.css 的代码模板如下：

【代码 1】{/* 代码 1 定义一个 class 选择符 myClass */
　　color: green;
}

第一段绿色内容。

第二段红色内容。

图 1.17　task1_2.html 的运行效果

【代码2】{/* 代码2定义一个id选择符myId */
    color: red;
}

task1_2.html 的代码模板如下：

```
<!DOCTYPE html PUBLIC "-//W3C//DTD HTML 4.01 Strict//EN"
"http://www.w3.org/TR/html4/strict.dtd">
<html>
    <head>
    <meta http-equiv="Content-Type" content="text/html; charset=UTF-8">
    <title>Insert title here</title>
    【代码3】<!-- 代码3引入myTask.css文件 -->
    </head>
    <body>
        <p【代码4】>第一段绿色内容。<!-- 代码4使用类选择符myClass -->
        <p【代码5】>第二段红色内容。<!-- 代码5使用id选择符myId -->
    </body>
</html>
```

3. 任务小结或知识扩展

使用外部样式文件时，需要注意样式文件的存放位置，即代码3中href的属性值。

4. 代码模板的参考答案

【代码1】：.myClass
【代码2】：#myId
【代码3】：<link rel="stylesheet" href="myTask.css" type="text/css" />
【代码4】：class="myClass"
【代码5】：id="myId"

1.2.4 实践环节

按照下面的步骤编写网页 practice1_2.html。

步骤1：使用DIV定义结构。

一个典型的版面分栏结构：页头、导航栏、内容、版权。结构代码如下：

```
<div id="header"></div>
<div id="navigator"></div>
<div id="content"></div>
<div id="footer"></div>
```

将这4个盒子装进一个更大的盒子，body中，代码如下：

```
<body>
    上面四行代码
</body>
```

步骤2：定义body的属性。

```
body {
    font-family: Arial, Helvetica, sans-serif;
```

```
    font-size: 12px;
    margin: 0px auto;
    height: auto;
    width: 800px;
    border: 1px solid #006633;
}
```

步骤3:定义页头(header)的属性。

```
#header {
    height: 100px;
    width: 800px;
    background-image: url(plane.jpg);
    background-repeat: no-repeat;
    margin: 0px 0px 3px 0px;
}
```

步骤4:定义导航栏(navigator)的属性。

```
#navigator { /*定义一个导航栏的长盒子*/
    height: 25px;
    width: 800px;
    font-size: 14px;
    list-style-type: none; /*让navigator这个大盒子下面的小盒子li列表样式不显示,这对于标准浏览器很重要*/
}

#navigator li {
    float: left; /*让li这些小盒子左对齐*/
}

#navigator li a {
    color: #000000;
    text-decoration: none; /*让li盒子里面的链接样式没有下划线*/
    padding-top: 4px;
    display: block; /*让li里面的链接块状呈现,就像一个按钮,而不必一定要点中链接文字才起作用*/
    width: 131px;
    height: 22px;
    text-align: center;
    background-color: #009966;
    margin-left: 2px;
}

#navigator li a:hover {
    background-color: #006633; /*鼠标移到链接盒子上面改变盒子背景色*/
    color: #FFFFFF;
}
```

步骤5:定义内容部分(content)的属性。

```
#content {
```

```css
    height: auto;
    width: 780px;
    line-height: 1.5em;
    padding: 10px;
}

#content p {
    text-indent: 2em;
}

#content h3 {
    font-size: 16px;
    margin: 10px;
}
```

步骤6：定义页脚(footer)的属性。

```css
#footer {
    height: 50px;
    width: 780px;
    line-height: 2em;
    text-align: center;
    background-color: #009966;
    padding: 10px;
}
```

步骤7：定义各标记的边界和填充(开头处)。

```css
* {
    margin: 0px;
    padding: 0px;
}
```

步骤8：结构代码如下。

```html
<body>
    <div id="header"></div>
    <div id="navigator">
        <ul id="navigator">
            <li><a href="#">首页</a></li>
            <li><a href="#">文章</a></li>
            <li><a href="#">相册</a></li>
            <li><a href="#">Blog</a></li>
            <li><a href="#">论坛</a></li>
            <li><a href="#">帮助</a></li>
        </ul>
    </div>
    <div id="content">
        <h3>前言</h3>
        <p>CSS是英文Cascading Style Sheet的缩写,又称为"层叠样式表",简称为样式表.它是W3C定义的标准,一种用来为结构化文档(如HTML文档)添加样式(字体、间距和背景等)的计算机语言</p>
```

```
            <h3>理解 DIV+CSS 布局</h3>
            <p>简单地说 DIV+CSS(DIV CSS)被称为"Web 标准"中常用术语之一. 首先认识 DIV 是用于搭
建 html 网页结构(框架)标签,再认识 CSS 是用于创建网页表现(样式/美化)样式表统称,通过 CSS 来
设置 div 标签样式,这一切常常称为 DIV+CSS
            </p>
        </div>
        <div id="footer">
            <p>关于|广告服务|招聘|客服中心|QQ留言|网站管理|会员登录|购物车</p>
            <p>Copyright ©清华大学出版社</p>
        </div>
</body>
```

页面运行效果如图 1.18 所示。

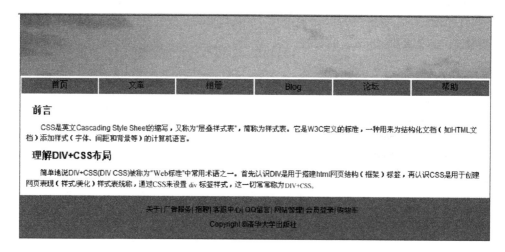

图 1.18　DIV+CSS 布局

1.3　JavaScript

JavaScript 同 CSS 一样,不需要使用复杂的工具来编写,可以使用文本编辑器或者 Web 开发工具来编写。

1.3.1　核心知识

1. 在网页中添加 JavaScript 的方法

1)嵌入使用

在网页代码中任何位置都可嵌入 JavaScript 代码,但建议嵌入 head 标记中。示例代码如下:

```
<html>
    <head>
        <meta http-equiv="Content-Type" content="text/html; charset=UTF-8">
        <title>JavaScript 嵌入学习</title>
        <script type="text/javascript">
```

```
            alert("第一次看到警告框很兴奋!");
        </script>
    </head>
    <body>
        好好学习 JavaScript 知识
    </body>
</html>
```

2) 引入使用

当多个页面使用相同的 JavaScript 代码时,可以将共用的代码保存在以 .js 为扩展名的文件中,然后在页面中使用 src 属性引入外部 js 文件。

myFirst.js 的代码模板如下:

```
alert("被引入页面中!");
```

引入外部 js 文件学习.html 的代码模板如下:

```
<html>
    <head>
        <meta http-equiv="Content-Type" content="text/html; charset=UTF-8">
        <title>引入外部 js 文件</title>
        <script type="text/javascript" src="myFirst.js" charset="GBK">
        </script>
    </head>
    <body>
        好好学习 JavaScript 知识
    </body>
</html>
```

2. JavaScript 基本语法

1) 变量

使用 var 可以声明任意类型的变量,例如:

```
var firstNumber = 10;
```

2) 类型转换

JavaScript 是弱类型语言,变量的类型对应于其值的类型。可以对不同类型的变量执行运算,解释器强制转换数据类型,然后进行处理。例如,数值与字符串相加,将数值强制转换为字符串;布尔值与字符串相加,将布尔值强制转换为字符串;数值与布尔值相加,将布尔值强制转换为数值。

字符串到数值的转换。例如,parseInt(s)将字符串转换为整数,parseFloat(s)将字符串转换为浮点数,Number(s)将字符串转换为数字。parseInt 和 parseFloat 方法只对 string 类型有效,且需要是数字开头的字符串。

3) 运算符

(1) 赋值运算符。赋值运算符的运算规则及说明如表 1.3 所示。

表 1.3　赋值运算符

运算符	示　例	说　明
＝	x＝y;	将变量 y 的值赋给 x
＋＝	x＋＝y;	x＝x＋y;
－＝	x－＝y;	x＝x－y;
＊＝	x＊＝y;	x＝x＊y;
/＝	x/＝y;	x＝x/y;

（2）数学运算符。数学运算符的运算规则及说明如表 1.4 所示。

表 1.4　数学运算符

运算符	示　例	说　明
＋	A＝5＋8　//结果是 13 A＝"5"＋8　//结果是"58"	如果操作数都是数字,执行加法运算,如果其中的操作数有字符串,会执行连接字符串的运算
－	A＝8－5	减法
＊	A＝8＊5	乘法
/	A＝8/5　//结果是 1.6	除法
％	10％3＝1	取余
++	++x 返回递增后的 x 值 x++返回递增前的 x 值	递增
－－	－－x 返回递减后的 x 值 x－－返回递减前的 x 值	递减
－	如果 a 等于 5,则－a＝－5	此运算符返回操作数的相反数

（3）逻辑运算符。逻辑运算符的运算规则及说明如表 1.5 所示。

表 1.5　逻辑运算符

运算符	示　例	说　明
&&	expr1 && expr2	逻辑与(表达式 1 错误,表达式 2 不再运算)
||	expr1 || expr2	逻辑或(表达式 1 正确,表达式 2 不再运算)
！	！expr	逻辑非

（4）typeof 运算符。对变量或值调用 typeof 运算符将返回对应的值,typeof 运算符的运算规则及说明如表 1.6 所示。

表 1.6　typeof 运算符

示　例	返回结果	说　明
typeof(true)	boolean	变量或值是 boolean 类型
typeof(300)	number	变量或值是 number 类型
typeof('abc')	string	变量或值是 string 类型
typeof(null)	object	变量或值是一种引用类型或 null 类型
typeof(f)	function	变量是一个函数

4) 注释

为了程序的可读性,以及便于日后代码修改和维护,可以在 JavaScript 程序里为代码写注释。在 JavaScript 程序里,用两个斜杠"//"来表示单行注释。多行注释用"/*"表示开始,"*/"表示结束。注释示例代码如下:

```
aGoodIdea = "Comment your code thoroughly."; //这是单行注释
/*
    这是多行注释 行一
    这是多行注释 行二
*/
```

5) 变量命名规则

以字母、下划线(_)或美元符号($)开头;余下的字符可以是下划线、美元符号或任何的字母、数字;不能有空格,大小写敏感;不能使用 JavaScript 中的关键字或保留字命名。

6) 部分保留字

break、delete、function、return、typeof、case、do、if、switch、var、catch、else、in、this、void、continue、false、instanceof、throw、while、finally、new、true、with、default、for、null、try

【例 1.6】 编写网页 example1_6.html,在网页中嵌入使用 JavaScript,JavaScript 程序的具体要求如下。

(1) 声明变量 firstNumber,并将 I am a String 赋值给该变量。
(2) 声明变量 secondNumber,并将 300 赋值给该变量。
(3) 声明变量 thirdNumber,并将 firstNumber+secondNumber 赋值给该变量。
(4) 使用 alert 弹出变量 thirdNumber 的值。
(5) 使用 typeof 运算符判断变量 firstNumber、secondNumber 以及 thirdNumber 的值类型,并使用 alert 分别弹出 typeof 的返回结果。

例 1.6 页面文件 example1_6.html 的代码如下:

```html
<!DOCTYPE html PUBLIC "-//W3C//DTD HTML 4.01 Transitional//EN"
"http://www.w3.org/TR/html4/loose.dtd">
<html>
    <head>
        <meta http-equiv="Content-Type" content="text/html; charset=UTF-8">
        <title>example1_6.html</title>
        <script type="text/javascript">
            var firstNumber = "I am a String";
            var secondNumber = 300;
            var thirdNumber = firstNumber + secondNumber;
            alert(thirdNumber);
            alert("firstNumber 的值类型: " + typeof(firstNumber));
            alert("secondNumber 的值类型: " + typeof(secondNumber));
            alert("thirdNumber 的值类型: " + typeof(thirdNumber));
        </script>
    </head>
    <body>
    </body>
</html>
```

3. 流程控制与函数

1）流程控制语句

(1) if 条件语句。

```
if(表达式){
    语句
}
```

或

```
if(表达式){
    语句
}else{
    语句
}
```

(2) switch 条件语句。

```
switch(表达式){
    case case1:
        语句
        break;
    case case2:
        语句
        break;
    …
    default:
        default 语句
}
```

(3) for 循环语句。

```
for(表达式1;表达式 2;表达式 3){
    语句
}
```

(4) while 循环语句。

```
while(表达式){
    语句
}
```

(5) do-while 循环语句。

```
do{
    语句
}while(表达式)
```

(6) break/continue 语句。break 语句让执行语句从循环语句或其他程序块中跳出。continue 语句让执行语句跳过本次循环的剩余语句进入下一次循环。

2）函数

常将完成某个功能的一组语句写成一个函数，函数的定义格式如下：

```
function 函数名([参数,参数]){
    函数体
}
```

function 是关键字,函数没有类型,参数也没有类型。例如:

```
function gogo(obj){
    document.write("函数没有类型,参数也没有类型");
}
```

3) arguments 对象

函数可以接受任意个数的参数,通过 arguments 对象来访问。示例代码如下:

```
function say(){
    if(arguments[1] != "你好"){
        alert(arguments[0]);
    }else{
        alert(arguments[1]);
    }
    alert(arguments.length);                //返回参数的个数
}
```

调用函数如下:

```
say("How are you?","你好");
```

4) 系统函数

JavaScript 提供了与任何对象无关的系统函数,这些函数不需要创建任何对象就可以直接使用。

(1) eval(字符串表达式)。

该函数的功能是返回字符串表达式的值,例如:

```
var test = eval("2+3");                     //test 的值为 5
```

(2) parseInt(字符串)。

该函数的功能是将以数字开头的字符串转换为整数,例如:

```
var test = parseInt("200.5abc");            //test 的值为 200
```

(3) parseFloat(字符串)。

该函数的功能是将以数字开头的字符串转换为浮点数,例如:

```
var test = parseFloat("200.5abc");          //test 的值为 200.5
```

(4) Number(字符串)。

该函数的功能是将数字字符串转换为数字,字符串中有非数字字符则返回 NaN。例如:

```
var test = Number("200.5abc");              //test 的值为 NaN
```

【例 1.7】 编写网页 example1_7.html,在网页中嵌入使用 JavaScript 程序打印出九九乘法表,网页运行效果如图 1.19 所示。

```
1*1=1
2*1=2  2*2=4
3*1=3  3*2=6  3*3=9
4*1=4  4*2=8  4*3=12 4*4=16
5*1=5  5*2=10 5*3=15 5*4=20 5*5=25
6*1=6  6*2=12 6*3=18 6*4=24 6*5=30 6*6=36
7*1=7  7*2=14 7*3=21 7*4=28 7*5=35 7*6=42 7*7=49
8*1=8  8*2=16 8*3=24 8*4=32 8*5=40 8*6=48 8*7=56 8*8=64
9*1=9  9*2=18 9*3=27 9*4=36 9*5=45 9*6=54 9*7=63 9*8=72 9*9=81
```

图 1.19 九九乘法表

例 1.7 页面文件 example1_7.html 的代码如下：

```html
<!DOCTYPE html PUBLIC "-//W3C//DTD HTML 4.01 Transitional//EN"
"http://www.w3.org/TR/html4/loose.dtd">
<html>
    <head>
        <meta http-equiv="Content-Type" content="text/html; charset=UTF-8">
        <title>example1_7.html</title>
        <script type="text/javascript">
        for(var i = 1; i <= 9; i++){
            for(var j = 1; j <= i; j++){
                document.write( i + "*" + j + "=" + i*j + " ");    //在页面输出
            }
            document.write("<br>");
        }
        </script>
    </head>
    <body>
    </body>
</html>
```

5) 函数调用

(1) 在链接中调用函数。用户单击链接时,即调用函数,格式如下:

```html
<a href="javascript:函数">…</a>
```

在链接中调用函数的示例代码如下:

```html
<html>
    <head>
        <meta http-equiv="Content-Type" content="text/html; charset=UTF-8">
        <title>链接调用函数</title>
        <script type="text/javascript">
        function gogo(){
            alert("被链接调用的函数");
        }
        </script>
    </head>
    <body>
        <a href="javascript:gogo()">链接调用函数</a>
```

```
    </body>
</html>
```

（2）由事件触发调用函数。触发事件调用函数，格式如下：

事件＝"函数"

触发事件调用函数的示例代码如下：

```
<html>
    <head>
        <meta http-equiv="Content-Type" content="text/html; charset=UTF-8">
        <title>事件触发调用函数</title>
        <script type="text/javascript">
        function gogo(param){
            alert(param);
        }
        </script>
    </head>
    <body>
        <form action="">
            <input type="button" value="点击我" onclick="gogo('0(∩_∩)0哈哈~')"/>
        </form>
    </body>
</html>
```

4. JavaScript 对象

一个 JavaScript 对象中可包含若干属性和方法。属性是描述对象的状态，对象用"."运算符访问属性。方法是描述对象的行为动作，对象用"."运算符调用方法。

1) 对象创建

使用 new 关键字来创建对象，例如：

```
var oStringObject = new String();
```

如果构造函数无参数，则不必加括号。

2) JavaScript 内部对象。

（1）数组（Array）对象。

① 创建数组。示例代码如下：

```
var myArray = new Array();              //新建一个长度为 0 的数组
var myArray = new Array(100);           //新建一个长度为 100 的数组
var myArray = new Array(1,2,3);         //新建一个指定长度的数组，并赋初值
```

数组长度不固定，赋值即可改变长度。数组的主要属性为 length，返回数组长度。

② 数组的常用方法。

reverse 方法：将 JavaScript 数组对象内容反转。

concat 方法：将两个或更多数组组合在一起，例如：

```
var newArray = tmpArray.concat(tmpArray);
```

join 方法：返回一个将数组所有元素用指定符号连在一起的字符串，例如：

```
var newString = tmpArray.join(".");
```

pop()方法：移除数组中的最后一个元素并返回该元素。
shift()方法：移除数组中的第一个元素并返回该元素。
slice 方法：返回数组的一部分，例如：

```
var newArray = tmpArray.slice(1,3);
```

③ 数组的使用。示例代码如下：

```
<script type="text/javascript">
    var myArray = new Array();
    for(var i = 0; i < 5; i++){
        myArray[i] = i;
    }
    for(var j = 0; j < myArray.length; j++){
        alert(myArray[j]);
    }
</script>
```

(2) 日期(Date)对象。Date 对象可以用来表示任意的日期和时间。

① 创建 Date 对象。必须使用 new 运算符创建一个 Date 对象。示例代码如下：

```
var date = new Date("July 8,2012");      //2012年7月8日
var date = new Date(2012,7,8);           //2012年7月8日
var date = new Date("2012/7/8") ;        //2012年7月8日
var date = new Date(Milliseconds);       //自1970年1月1日以来的毫秒数创建的日期对象
var date = new Date();                   //当前系统的时间对象
```

② 获取日期的时间方法。

getYear()：返回年数。

getMonth()：返回当月号数（比实际值小 1）。

getDate()：返回当日号数。

getDay()：返回星期几(0 表示周日)。

getHours()：返回小时数。

getMinutes()：返回分钟数。

getSeconds()：返回秒数。

getTime()：返回毫秒数。

③ 设置日期和时间的方法。

setYear()：设置年数。

setMonth()：设置当月号数(set6 表示 7 月)。

setDate()：设置当日号数。

setHours()：设置小时数。

setMinutes()：设置分钟数。

setSeconds()：设置秒数。

setTime()：设置毫秒数。

④ Date 对象的使用。示例代码如下：

```
<script type="text/javascript">
    var date = new Date("2050/12/25");
    document.write("2050 的圣诞节是星期" + date.getDay() + "<br>");
    var datenow = new Date();        //得到当前日期对象
    var mills = date - datenow;      //两个 Date 对象相减得到两个日期的时间间隔(单位是毫秒)
    document.write("2050 的圣诞节距离现在还有" + mills + "毫秒<br>");
</script>
```

(3) String 对象。

① 创建 String 对象。示例代码如下：

```
var firstString = "This is a string";
var secondString = new String("This is a string");
```

String 对象的主要属性为 length，返回字符串的长度。

② String 对象的常用方法。

charAt(i)：返回指定索引位置处的字符，索引从 0 开始。

concat(str)：连接字符串。

indexOf(str)：返回 String 对象内第一次出现子字符串的字符位置（从左到右查找）。

lastIndexOf(str)：返回 String 对象中子字符串最后出现的位置。

replace(str1,str2)：返回将 str1 替换为 str2 后的字符串。

split(separator, limit)：将字符串以 separator 作为分割符切割成多个子字符串，并将它们作为一个数组返回；如果有 limit 参数则返回数组的 limit 个元素。

substring(start,end)：返回一个指定位置之间的子字符串，不包括 end。

toLowerCase()：返回一个字符串，字符串中的字母被转换为小写字母。

toUpperCase()：返回一个字符串，字符串中的字母被转换为大写字母。

③ String 对象的使用。示例代码如下：

```
<script type="text/javascript">
    var firstString = "This is a string";
    for(var i = 0; i < firstString.length; i++){
        alert(firstString.charAt(i));
    }
</script>
```

(4) Math 对象。Math 对象是一个全局对象，使用时不需要创建。

① Math 对象的属性。

LN10：10 的自然对数。

LN2：2 的对数。

PI：圆周率。

SQRT1_2：1/2 的平方根。

SQRT2：2 的平方根。

② Math 对象的常用方法。

abs(x)：返回 x 的绝对值。
ceil(x)：返回大于等于 x 的最小整数。
floor(x)：返回小于等于 x 的最大整数。
round(x)：舍入到最近整数。
sqrt(x)：返回 x 的平方根。
random()：返回 0~1 的随机数。
③ Math 对象的使用。示例代码如下：

```
<script type="text/javascript">
    alert(Math.SQRT2);
    alert(Math.random());
</script>
```

【例 1.8】 编写网页 example1_8.html，在网页中定义一个 JavaScript 函数，功能是去除字符串开头及末尾的空格，并使用超链接来调用该函数。

例 1.8 页面文件 example1_8.html 的代码如下：

```
<!DOCTYPE html PUBLIC "-//W3C//DTD HTML 4.01 Transitional//EN"
"http://www.w3.org/TR/html4/loose.dtd">
<html>
    <head>
        <title>example1_8.html</title>
        <script type="text/javascript">
            function print99(x){
                while((x.length>0) && (x.charAt(0) == ' '))
                    x = x.substring(1,x.length);
                while(x.length>0&&(x.charAt(x.length-1) == ' '))
                    x = x.substring(0,x.length-1);
                alert("Kill=" + x + "==")
                return x;
            }
        </script>
    </head>
    <body>
        <a href="javascript:print99(' abc def ')">点击我啊!</a>
    </body>
</html>
```

5. JavaScript 对象模型

1) 浏览器对象模型

浏览器对象是提供独立于内容而与浏览器窗口进行交互的对象。浏览器对象模型如图 1.20 所示。

2) window 对象

(1) 打开新窗口。使用 open() 方法可打开一个新窗口，示例代码如下：

```
var winObj = open("target.html", "target_1","width=500,height=300,scrollbars=no");
```

第 1 章 Web 前端基础

图 1.20 浏览器对象模型

open()方法有 3 个参数：第一个参数代表要载入新窗口页面的 URL，第二个参数代表新窗口的名称，第三个参数代表新窗口的属性，多个属性间用逗号隔开。

（2）对话框（与用户交互）方法。

① alert()。该方法的功能是弹出一个提示框。示例代码如下：

```
<script type="text/javascript">
    alert("请单击确定按钮!");
</script>
```

运行效果如图 1.21 所示。

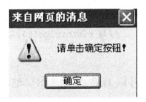

图 1.21 具有"确定"按钮的对话框

② prompt(message,defaultText)。该方法的功能是弹出可以输入信息的文本框，第一个参数代表用户输入信息的提示，第二个参数代表文本框的默认值。示例代码如下：

```
<script type="text/javascript">
    prompt("What's your name?","chenheng");
</script>
```

运行效果如图 1.22 所示。

③ confirm(message)。该方法的功能是弹出对话框，提示确认信息。示例代码如下：

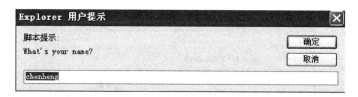

图 1.22 用户输入对话框

```
<script type="text/javascript">
    if(confirm("真的删除吗?")){
        //单击"确定"按钮后的操作
    }else{
        //单击"取消"按钮后的操作
    }
</script>
```

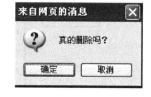

图 1.23 确认对话框

运行效果如图 1.23 所示。

3) history 对象

history 对象记录着浏览器所浏览过的每一个页面,这些页面组成了一个历史记录列表。它有以下 3 个主要方法。

forward():将历史记录向前移动一个页面。

back():将历史记录向后移动一个页面,在网页中经常使用该方法提供一个"返回"的功能。

go():转向历史记录中指定的地址,使用此方法需要一个参数,参数可以是正负整数或字符串。如果参数是字符串,那么浏览器就会搜索列表,找到最接近当前页面位置且 URL 地址中含有此字符串的页面,然后转到该页面。

history 对象的使用,示例代码如下:

```
history.go(-3);              //向后返回3个访问过的页面
histroy.go(3);               //向前返回3个访问过的页面
history.back();              //与 history.go(-1);功能相同
history.forward();           //与 history.go(1);功能相同
```

4) location 对象

window 对象的 location 属性可以直接改变 URL 地址:

```
window.location = "http://www.baidu.com";
```

或

```
location = "http://www.baidu.com";
```

还可以使用 location 对象的 href 属性或 replace(URL)方法改变 URL 地址:

```
location.href = "http://www.baidu.com";
```

或

```
location.replace("http://www.baidu.com");
```

可以使用 location 对象的 href 属性清空页面：

```
location.href = "about:blank";    //清空页面
```

5) document 对象

(1) forms 集合。在同一个页面上有多个表单，通过 document.forms[] 数组获得这些表单对象要比使用表单名称方便得多。

(2) getElementById(id)方法。该方法的功能是获得指定 id 值的表单控件对象。

(3) getElementsByName(name)方法。该方法的功能是获得指定 name 值的表单控件对象，返回的是对象数组。

(4) 获取表单对象的方法。获取表单对象的方法如下：

```
document.forms[0];              //通过 forms 对象的索引
document.forms["myForm"];       //通过 forms 对象和表单名称
document.myForm;                //通过表单名称
```

(5) document 对象的使用。document 对象的使用，示例代码如下：

```html
<html>
    <head>
        <title>The Document Object</title>
        <script type="text/javascript">
            function testMethod() {
                //(1)通过表单获取控件
                var name_1 = document.forms[0].nameTest;
                alert("name_1 = " + name_1.value);
                var name_2 = document.forms["myForm"].nameTest;
                alert("name_2 = " + name_2.value);
                var name_3 = document.myForm.nameTest;
                alert("name_3 = " + name_3.value);
                var name_4 = document.myForm.elements["nameTest"];
                alert("name_4 = " + name_4.value);
                //(2)通过 name 获取控件
                var name_5 = document.getElementsByName("nameTest");
                alert("name_5[0] = " + name_5[0].value);
                //(3)通过 id 获取控件
                var id_1 = document.getElementById("idTest");
                alert("id_1 = " + id_1.value);
            }
        </script>
    </head>
    <body>
        <form name="myForm">
            通过 name 获取：<input type="text" name="nameTest" value="firstTextValue"><br>
            <br>通过 id 获取：<input type="text" id="idTest" value="secondTextValue">
            <br><input type="button" value="Click Me" onclick="testMethod()">
        </form>
    </body>
</html>
```

6. 操作 HTML

1) 事件处理

通常将鼠标或热键的动作称为事件。由鼠标或热键引发的一连串程序的动作，称为事件驱动。对事件进行处理的程序或函数称为事件处理程序。

（1）窗口或页面的事件处理。窗口或页面的事件处理如表 1.7 所示。

表 1.7　窗口或页面的事件处理

事　　件	说　　明
onBlur	当前元素失去焦点时触发
onFocus	当某个元素获得焦点时触发
onLoad	页面内容完成装载时触发
onUnload	当前页面被退出或重置时触发

（2）键盘或鼠标的事件处理。键盘或鼠标的事件处理如表 1.8 所示。

表 1.8　键盘或鼠标的事件处理

事　　件	说　　明
onClick	当鼠标单击时触发
onDblClick	当鼠标双击时触发
onMouseDown	当按下鼠标时触发
onMouseMove	当鼠标移动时触发
onMouseOut	当鼠标离开某对象范围时触发
onMouseOver	当鼠标移动到某对象范围的上方时触发
onMouseUp	当鼠标按下后松开鼠标时触发
onKeyPress	当键盘上某个键被按并且释放时触发
onKeyDown	当键盘上某个键被按时触发
onKeyUp	当键盘上某个键被释放时触发

（3）表单元素的事件处理。表单元素的事件处理如表 1.9 所示。

表 1.9　表单元素的事件处理

表 单 元 素	主 要 事 件
button(按钮)	onClick onBlur onFocus
checkbox(复选框)	onClick onBlur onFocus
file(上传文件)	onClick onBlur onFocus
password(密码框)	onBlur onFocus onSelect
radio(单选按钮)	onClick onBlur onFocus
select(列表)	onFocus onBlur onChange
text(文本框)	onClick onBlur onFocus onChange
textarea(文本区)	onClick onBlur onFocus onChange

(4) 事件处理模型。在 JavaScript 中对事件的处理程序通常由函数完成,事件＝"函数名",例如:

```html
<html>
    <head>
        <title>The Document Object</title>
        <script type="text/javascript">
            function testMethod() {
                alert("事件处理模型");
            }
        </script>
    </head>
    <body>
        <form name="myForm">
            <input type="button" value="Click Me" onclick="testMethod()">
        </form>
    </body>
</html>
```

2) 表单元素

(1) 表单元素的通用属性与方法。

form 属性:获取该表单控件所属的表单对象。

name 属性:获取或设置表单控件的名称。

type 属性:获取表单控件的类型。

value 属性:获取和设置表单控件的值。

focus 方法:让表单控件对象获得焦点。

blur 方法:让表单控件对象失去焦点。

表单元素的通用属性及方法的示例代码如下:

```html
<html>
    <head>
        <title>form</title>
        <script type="text/javascript">
            //显示属性
            function test() {
                var text_name = document.myForm.text_name;
                alert("text_name.form = " + text_name.form.name
                    + "\ntext_name.name = " + text_name.name
                    + "\ntext_name.type = " + text_name.type
                    + "\ntext_name.value = " + text_name.value
                    + "\ntext_name.defaultValue = " + text_name.defaultValue);
            }
            //获得焦点
            function do_focus() {
                document.myForm.text_name.focus();
            }
            //失去焦点
            function do_blur() {
                document.myForm.text_name.blur();
```

```
            }
        </script>
    </head>
    <body>
        <form name = "myForm">
            < input type = "text" name = "text_name" value = "textValue"><br>
            < input type = "button" name = "button_1" value = "显示属性" onclick = "test()"><br>
            < input type = "button" name = "button_2" value = "获得焦点" onclick = "do_focus()">
            <br>
            < input type = "button" name = "button_3" value = "失去焦点" onclick = "do_blur()">
        </form>
    </body>
</html>
```

(2) 文本框。

value 属性：获得文本框的值，值是字符串类型。

defaultValue 属性：获得文本框的默认值，值是字符串类型。

readonly 属性：只读，文本框中的内容不能修改。

focus 方法：获得焦点，即获得鼠标光标。

blur 方法：失去焦点。

select 方法：选中文本框内容，突出显示输入区域。

文本框的示例代码如下：

```
<html>
    <head>
        <title>文本框求和</title>
        < script type = "text/javascript">
        function add(){
            var sum = 0;
            var text_1Value = document.forms[0].text_1.value;
            var text_2Value = document.forms[0].text_2.value;
            sum = Number(text_1Value) + Number(text_2Value);
            document.forms[0].text_3.value = sum;
        }
        </script>
    </head>
    <body>
        < form name = "form1" method = "post" action = "">
            < input type = "text" name = "text_1" value = ""><br>
            < input type = "text" name = "text_2" value = ""><br>
            < input type = "text" name = "text_3" value = ""><br>
            < input type = "button" value = "求和" onclick = "add()">
        </form>
    </body>
</html>
```

(3) 复选框。

checked 属性：复选框是否被选中，选中为 true，未选中为 false。

value 属性：设置或获取复选框的值。

复选框的示例代码如下：

```html
<html>
    <head>
        <title>复选框反选</title>
        <script type="text/javascript">
            function unSelect() {
                var n = document.forms[0].check.length;    //得到复选框的个数
                for ( var i = 0; i < n; i++) {
                    if (document.forms[0].check[i].checked) {
                        document.forms[0].check[i].checked = false;
                    } else {
                        document.forms[0].check[i].checked = true;
                    }
                }
            }
        </script>
    </head>
    <body>
        <form action="">
            <input type="checkbox" name="check" value="0" />aaa<br>
            <input type="checkbox" name="check" value="1" />bbb<br>
            <input type="checkbox" name="check" value="2" />ccc<br>
            <input type="checkbox" name="check" value="3" />ddd<br>
            <input type="checkbox" name="check" value="4" />eee<br>
            <input type="checkbox" name="check" value="5" />fff<br>
            <input type="button" value="反选" onclick="unSelect()">
        </form>
    </body>
</html>
```

(4) 单选按钮。

checked 属性：单选按钮是否被选中，选中为 true，未选中为 false。

value 属性：设置或获取单选按钮的值。

单选按钮的示例代码如下：

```html
<html>
    <head>
        <title>弹出单选按钮的值</title>
        <script type="text/javascript">
            function gg() {
                var n = document.forms[0].sex.length;
                for ( var i = 0; i < n; i++) {
                    if (document.forms[0].sex[i].checked) {
                        alert(document.forms[0].sex[i].value);
                    }
                }
            }
        </script>
    </head>
```

```html
<body>
    <form action = "">
        <input type = "radio" name = "sex" value = "male" onclick = "gg()" />男
        <input type = "radio" name = "sex" value = "female" onclick = "gg()" />女
    </form>
</body>
</html>
```

（5）下拉列表。

length 属性：选项个数。

selectedIndex 属性：当前被选中选项的索引。

options 属性：所有的选项组成一个数组，options 表示整个选项数组，第一个选项即为 options[0]，第二个即为 options[1]，其他以此类推。

option 的 value 属性：option 标记中 value 所指定的值。

option 的 text 属性：显示于界面中的文本，即<option>…</option>之间的部分。

增加选项：每个选项都是一个 option 对象，可以创建 option 对象，然后添加到 select 的末尾。例如：

```
select.options[select.length] = new Option(text,value);
```

下拉列表的示例代码如下：

```html
<html>
    <head>
    <title>下拉列表的应用</title>
        <script type = "text/javascript">
            function gg() {
                var opObject = document.forms[0].elements["cities"];  //获得列表对象
                for ( var i = 0; i < opObject.options.length; i++) {   //使用 options 属性
                    if (opObject.options[i].selected) {                //找到被选中的选项
                        alert(opObject.options[i].value);              //弹出选中选项的值
                        alert(opObject.options[i].text);               //弹出选中选项的文本
                    }
                }
                //为列表新增选项
                opObject.options[opObject.length] = new Option("新增" + 1, "new" + 1);
            }
        </script>
    </head>
    <body>
        <form action = "">
            <select name = "cities" onchange = "gg()">
                <option value = "dalian">大连</option>
                <option value = "beijing">北京</option>
                <option value = "shanghai">上海</option>
                <option value = "guangzhou">广州</option>
            </select>
        </form>
    </body>
```

</html>

7. 表单验证

用户在表单中输入的内容提交到服务器之前,在客户端利用表单控件产生的事件,运用JavaScript,验证用户输入数据的有效性。

1) 正则表达式

表单验证常用的正则表达式如下。

验证邮政编码:/^\d{6}$/
验证身份证号码:/^\d{15}$|^\d{18}$|^\d{17}[xX]$/
验证电子邮箱地址:/^\w+((-\w+)|(\.\w+))*\@[A-Za-z0-9]+((\.|-)[A-Za-z0-9]+)*\.[A-Za-z0-9]+$/
验证数字或英文字母:/^[a-z0-9]+$/
验证日期格式:
/((^((1[8-9]\d{2})|([2-9]\d{3}))([-\/\._])(10|12|0?[13578])([-\/\._])(3[01]|[12][0-9]|0?[1-9])$)|(^((1[8-9]\d{2})|([2-9]\d{3}))([-\/\._])(11|0?[469])([-\/\._])(30|[12][0-9]|0?[1-9])$)|(^((1[8-9]\d{2})|([2-9]\d{3}))([-\/\._])(0?2)([-\/\._])(2[0-8]|1[0-9]|0?[1-9])$)|(^([2468][048]00)([-\/\._])(0?2)([-\/\._])(29)$)|(^([3579][26]00)([-\/\._])(0?2)([-\/\._])(29)$)|(^([1][89][0][48])([-\/\._])(0?2)([-\/\._])(29)$)|(^([2-9][0-9][0][48])([-\/\._])(0?2)([-\/\._])(29)$)|(^([1][89][2468][048])([-\/\._])(0?2)([-\/\._])(29)$)|(^([2-9][0-9][2468][048])([-\/\._])(0?2)([-\/\._])(29)$)|(^([1][89][13579][26])([-\/\._])(0?2)([-\/\._])(29)$)|(^([2-9][0-9][13579][26])([-\/\._])(0?2)([-\/\._])(29)$))/

正则表达式的应用示例代码如下:

```
<html>
    <head>
        <title>正则表达的应用</title>
        <script type="text/javascript">
        //验证E-mail
        function checkEmail(){
            var exp = /^\w+((-\w+)|(\.\w+))*\@[A-Za-z0-9]+((\.|-)[A-Za-z0-9]+)*\.[A-Za-z0-9]+$/;
            if(!exp.test(document.forms[0].email.value)){
                alert("E-mail 格式错误!");
                document.forms[0].email.focus();
                return false;
            }else{
                alert("E-mail 格式正确!");
                return true;
            }
        }
        </script>
    </head>
    <body>
        <form action="">
            Mail: <input type="text" name="email" /><br>
            <input type="button" value="提交" onclick="return checkEmail()" />
        </form>
```

```
    </body>
</html>
```

2）表单验证实例

验证密码域：不能为空，长度要大于等于6，只能是字母或数字。代码如下：

```
<html>
    <head>
        <title>表单验证实例</title>
        <script type="text/javascript">
        function valid(form){
            //验证非空
            if(form.pass.value.length == 0){
                alert("Please enter a password");
                form.pass.focus();
                return false;
            }
            //验证长度
            if(form.pass.value.length < 6){
                alert("Password must be at least 6 characters");
                form.pass.focus();
                return false;
            }
            var exp = /^[a-z0-9]+$/;
            //验证格式
            if(!exp.test(form.pass.value)){
                alert("Password contains illegal characters");
                form.pass.focus();
                return false;
            }
            alert("OK password");
            return true;
        }
        </script>
    </head>
    <body>
        <form action="">
            Enter your password:
            <input name="pass" type="password"/>
            <input type="button" value="submit password" onclick="return valid(this.form)"/>
        </form>
    </body>
</html>
```

1.3.2 能力目标

通过本节的学习，掌握JavaScript的基本语法、事件处理和常用对象，能够编写基本的表单验证程序。

1.3.3 任务驱动

1. 任务的主要内容

首先,编写一个 JS 文件 myTask3.js,在该 JS 文件中定义一个判断文本框值非空的函数;然后,再编写一个 HTML 文件 task1_3.html,在该 HTML 文件中使用 myTask3.js 的函数验证表单中文本框的值是否输入。task1_3.html 和 myTask3.js 在同一个目录下。task1_3.html 的运行效果如图 1.24 所示。

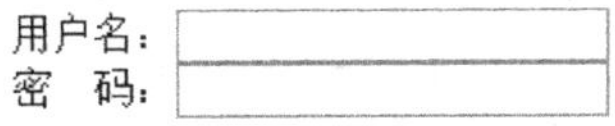

2. 任务的代码模板

myTask3.js 的代码模板如下:

用户名:
密　码:

图 1.24　task1_3.html 的运行效果

```javascript
function isNull(parameterObject,parameterName) {
    var parameterValue = parameterObject.value;
    if(parameterValue == ""){
        alert(parameterName + "不能为空");
        return false;
    }
    return true;
}
```

task1_3.html 的代码模板如下:

```html
<!DOCTYPE html PUBLIC "-//W3C//DTD HTML 4.01 Strict//EN"
"http://www.w3.org/TR/html4/strict.dtd">
<html>
    <head>
        <meta http-equiv="Content-Type" content="text/html; charset=UTF-8">
        <title>Insert title here</title>
    </head>
    <body>
        <form action="">
            用户名:<input type="text" name="username"【代码 1】="return isNull(this,this.name)"><br>
            <!-- 代码 1 使用 onblur 事件调用函数 isNull 判断文本框的值是否输入 -->
            密 码:<input type="password" name="userpwd"【代码 2】><br>
            <!-- 代码 2 使用 onblur 事件调用函数 isNull 判断密码框的值是否输入 -->
        </form>
    </body>
【代码 3】<!-- 代码 3 使用 script 标记引入 myTask3.js 文件 -->
</html>
```

3. 任务小结或知识扩展

当多个页面使用相同的 JavaScript 代码时,建议将共用的代码保存在以.js 为扩展名的文件中,然后在页面中使用 script 标记引入 JS 文件。但需要注意的是,引入 JS 文件可能会延迟页面的显示。所以,在不影响使用的情况下,页面最后引入 JS 文件。

4. 代码模板的参考答案

【代码 1】：onblur

【代码 2】：onblur = "return isNull(this,this.name)"

【代码 3】：＜script type = "text/javascript" src = "myTask3.js"＞＜/script＞

1.3.4 实践环节

制作一个用户注册页面 practice1_3.html，具体要求如下。

(1) 有常用的登录账号、密码、确认密码、姓名、身份证号码(只考虑 18 位的身份证)、出生年月日、住址、邮编、E-mail 等输入区域(自己设定)。

(2) 自己设定验证规则，在提交时检验是否符合要求，提示非法的输入，并将焦点返回要输入的控件对象。

(3) 根据出生年月日判断身份证号码是否合法(只考虑 18 位的身份证)。

(4) 其他验证：登录账号只能是字母或数字且以字母开头；密码要在 8 位以上且需要有字母和数字之外的字符；出生年月日的格式为 yyyy-mm-dd；邮编为 6 位数字；E-mail 的基本格式验证。

1.4 小　　结

本章简要介绍了 HTML、CSS 和 JavaScript，更多内容请参考其他教材或访问 http://www.w3school.com.cn 网站。

习　题　1

1. 以下标记中，用于设置页面标题的是(　　)。
   A. title 　　 B. caption 　　 C. head 　　 D. html
2. 若要设计网页的背景图形为 bg.png，以下标记中，正确的是(　　)。
   A. ＜body background＝"bg.png"＞
   B. ＜body bground＝"bg.png"＞
   C. ＜body image＝"bg.png"＞
   D. ＜body bgcolor＝"bg.png"＞
3. 关于 HTML 文件的说法正确的是(　　)。
   A. HTML 标记都必须配对使用
   B. 在＜title＞和＜/title＞标签之间的是头信息
   C. HTML 标签是区分大小写的，＜B＞与＜b＞表示的意思是不一样的
   D. ＜!－－＞标记是注释标记
4. 下列表示的不是按钮的是(　　)。
   A. type＝"submit" 　　 B. type＝"reset"
   C. type＝"image" 　　 D. type＝"button"

5. 若要产生一个 4 行 30 列的多行文本域,以下方法中,正确的是(　　)。
   A. <input type="text" rows="4" cols="30" name="txtintrol">
   B. <textarea rows="4" cols="30" name="txtintro">
   C. <textarea rows="4" cols="30" name="txtintro"></textarea>
   D. <textarea rows="30" cols="4" name="txtintro"></textarea>
6. 用于设置文本框显示宽度的属性是(　　)。
   A. size   B. maxlength   C. value   D. length
7. 下面对表单的说法错误的是(　　)。
   A. 表单在 Web 页面中用来给访问者填写信息,从而能采集客户端信息,使页面具有交互信息的功能
   B. 当用户填写完信息后单击普通按钮做提交(submit)操作
   C. 一个表单用<form></form>标记来创建
   D. action 属性的值是指处理程序的程序名(包括网络路径、网址或相对路径)
8. 下面说法错误的是(　　)。
   A. 在 HTML 语言中,input 标记具有重要的地位,它能够将浏览器中的控件加载到 HTML 文档中,该标记既有开始标记,又有结束标记
   B. <input type="text">是设定一个单行的文本输入区域
   C. size 属性指定控件宽度,表示该文本输入框所能显示的最大字符数
   D. maxlength 属性表示该文本输入框允许用户输入的最大字符数
9. 下面对于按钮的说法中正确的是(　　)。
   A. 按钮可分为普通按钮、提交按钮和重置按钮
   B. <input type="button">表示这是一个提交到服务器的按钮
   C. <input type="reset">表示这是一个普通按钮
   D. name 属性用来指定按钮页面显示的名称
10. 下面说法中错误的是(　　)。
    A. <input type="checkbox " checked>,其中 checked 属性用来设置该复选框默认时是否被选中
    B. <input type="hidden">表示一个隐藏区域。用户可以在其中输入某些要传送的信息
    C. <input type="password">表示这是一个密码区域。当用户输入密码时,区域内将会显示"＊"号
    D. <input type="radio">表示这是一个单选按钮
11. 下列说法中错误的是(　　)。
    A. <select></select>标记对用来创建一个菜单下拉列表框
    B. 下拉列表框中 multiple 属性不用赋值,直接加入标记中即可使用,加入此属性后列表框就可多选
    C. <option>标记用来指定列表框中的一个选项
    D. 不可以设定输入多行的文本区的大小
12. 下面对样式表的说法中错误的是(　　)。

A. CSS 就是 Cascading Style Sheets,中文翻译为"层叠样式表",简称样式表

B. 将 CSS 指定的格式加入 HTML 中的方法有两种

C. 外部链接 CSS 时需要在 HTML 文件里加一个超链接,链接到外部的 CSS 文档

D. 内定义 CSS 时需要在 HTML 文件内的<head>…</head>标记之间,加一段 CSS 的描述内容

13. 下面对样式表的说法中错误的是(　　)。

    A. CSS 的定义是由 3 个部分构成:选择符(selector)、属性(properties)和属性的取值(value)

    B. 选择符可以是多种形式,一般是要定义样式的 HTML 标记,可以通过此方法定义它的属性和值,属性和值要用逗号隔开

    C. CSS 可以定义字体属性

    D. CSS 可以定义颜色和背景属性

14. 下列说法中错误的是(　　)。

    A. p{font-family:"sans serif"}定义段落字体为 sans serif

    B. p{text-align:center;color:red}定义段落居中排列;并且段落中的文字为红色

    C. background-image 属性用来设置背景图片

    D. background-color 属性用来设置背景颜色

15. (　　)对象表示浏览器的窗口,可用于检索关于该窗口状态的信息。

    A. document　　B. window　　C. location　　D. history

16. (　　)对象表示给定浏览器窗口中的 HTML 文档,用于检索关于文档的信息。

    A. document　　B. window　　C. screen　　D. history

17. (　　)方法要求窗口显示刚刚访问的前一个页面。

    A. back()　　B. go()　　C. display()　　D. view()

18. 有关变量的命名规则,下列说法中错误的是(　　)。

    A. 以字母、下划线(_)或美元符号($)开头

    B. 首字符之外的字符可以是下划线、美元符号或任何的字母、数字

    C. 不能有空格,不区分大小写

    D. 不能使用 JavaScript 中的关键字或保留字命名

19. 单击按钮触发的事件是(　　)。

    A. onClick　　B. onFocus　　C. onChange　　D. onLoad

20. 页面加载时产生的事件是(　　)。

    A. onClick　　B. onFocus　　C. onChange　　D. onLoad

21. 请用 HTML 编写网页版个人简历。具体要求如下。

(1) 页面内容以 DIV+CSS 形式体现。

(2) 信息内容包括个人基础信息、近期照片、学习经历、兴趣爱好、求职意向。

第 2 章 JSP 简介及开发环境的构建

主要内容

(1) 开发环境的构建。
(2) 使用 Eclipse 开发 Web 应用。

JSP 是 Java Server Pages(Java 服务器页面)的简称,是基于 Java 语言的一种 Web 应用开发技术,是由 Sun 公司倡导,多家公司共同参与建立的一种动态网页技术标准。

在学习 JSP 之前,读者应具有 Java 语言基础以及 HTML 语言方面的知识。本章 2.2 节通过一个简单的 Web 应用介绍 JSP 项目开发的基本步骤,这些基本步骤对后续章节的学习是极其重要的。

2.1 构建开发环境

2.1.1 核心知识

所谓"工欲善其事,必先利其器",在开发 JSP 应用程序时,需要事先构建其开发环境。

1. Java 开发工具包

JSP 引擎需要 Java 语言的核心库和相应编译器,在安装 JSP 引擎之前,需要安装 Java 标准版(Java SE)提供的开发工具包 JDK。登录 http://www.oracle.com/technetwork/java,在 Software Downloads 列表中选择 Java SE 提供的 JDK,例如 Java Platform (JDK) 7u3。根据操作系统的位数,下载相应的 JDK,例如 32 位的系统使用 32 位的 JDK。本书采用的 JDK 是 jdk-7u3-windows-x64.exe。

2. JSP 引擎

运行包含 JSP 页面的 Web 项目还需要一个支持 JSP 的 Web 服务软件,该软件也称作 JSP 引擎或 JSP 容器,通常将安装 JSP 引擎的计算机称为一个支持 JSP 的 Web 服务器。目前,比较常用的 JSP 引擎包括 Tomcat、JRun、Resin、WebSphere、WebLogic 等,本书采用的是 Tomcat。

登录 Apache 软件基金会的官方网站 http://jakarta.Apache.org/tomcat,下载 Tomcat 7.0 的免安装版(apache-tomcat-7.0.57.zip)。登录网站后,首先在 Download 列表中选择 Tomcat 7.0,然后在 Binary Distributions 的 Core 中选择 zip 即可。

3. Eclipse

为了提高开发效率,通常还需要安装 IDE(集成开发环境)工具,本书使用的 IDE 工具是 Eclipse。Eclipse 是一个可用于开发 JSP 程序的 IDE 工具。登录 http://www.eclipse.org,在 Downloads 列表中选择 Eclipse IDE for Java EE Developers,然后在 Download Links 列表中,根据操作系统的位数,下载相应的 Eclipse。本书采用的 Eclipse 是 eclipse-jee-luna-SR1-win32-x86_64.zip。

2.1.2 能力目标

安装与配置 JSP 的运行环境。

2.1.3 任务驱动

任务的主要内容如下。
(1) JDK 的安装与配置。
(2) Tomcat 的安装与启动。
(3) Eclipse 的安装与配置。

1. JDK 的安装与配置

1) 安装 JDK

双击下载后的 jdk-7u3-windows-x64.exe 文件图标,出现安装向导界面,选择接受软件安装协议。建议采用默认的安装路径 C:\Program Files\Java\jdk1.7.0_03。需要注意的是,在安装 JDK 的过程中,JDK 还额外提供一个 Java 运行环境 JRE(Java Runtime Environment),并提示是否修改 JRE 默认的安装路径 C:\program Files\Java\JRE7,建议采用该默认的安装路径。

2) 配置系统环境变量

安装 JDK 平台之后需要进行几个系统环境变量的设置。

(1) 配置系统环境变量 Java_Home。在 Windows 7 系统下,右击"计算机"图标,在弹出的菜单中选择"属性"命令,单击"高级系统设置"按钮,单击"环境变量"按钮,在"新建系统变量"对话框的"变量名"文本框中输入 Java_Home,在"变量值"文本框中输入 C:\Program Files\Java\jdk1.7.0_03,如图 2.1 所示。

图 2.1 "新建系统变量"对话框

(2) 编辑系统环境变量 Path。双击系统变量 Path 进行编辑操作,在"变量值"文本框中的最前面追加"C:\Program Files\Java\jdk1.7.0_03\bin;",如图 2.2 所示。

图 2.2 "编辑系统变量"对话框

2. Tomcat 的安装与启动

安装 Tomcat 之前需要事先安装 JDK。将下载的 apache-tomcat-7.0.57.zip 解压到磁盘的某个分区中，比如解压到 D:\，解压缩后将出现如图 2.3 所示的目录结构。

图 2.3　Tomcat 目录结构

执行 Tomcat 根目录中 bin 文件夹中的 startup.bat 文件来启动 Tomcat 服务器。执行 startup.bat 文件启动 Tomcat 服务器会占用一个 MS-DOS 窗口，出现如图 2.4 所示的界面，如果关闭当前 MS-DOS 窗口将关闭 Tomcat 服务器。

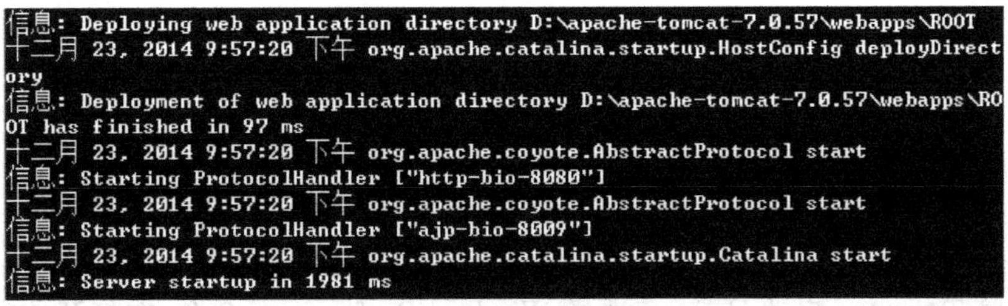

图 2.4　执行 startup.bat 文件启动 Tomcat 服务器

Tomcat 服务器启动后，在浏览器的地址栏中输入 http://localhost:8080，将出现如图 2.5 所示的 Tomcat 测试页面。

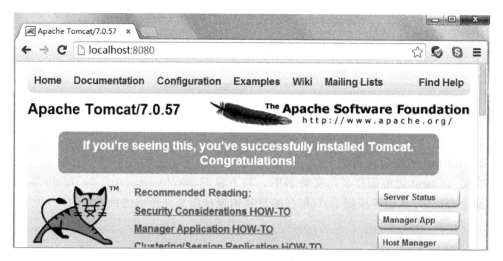

图 2.5　Tomcat 测试页面

3. Eclipse 的安装与配置

使用 Eclipse 开发 JSP 程序之前，需要对 JDK、Tomcat 和 Eclipse 进行一些必要的配置。因此，在安装 Eclipse 之前，应该事先安装 JDK 和 Tomcat。

1）安装 Eclipse

Eclipse 下载完成后，解压到自己设置的路径下，即可完成安装。Eclipse 安装后，双击 Eclipse 安装目录下的 eclipse.exe 文件，启动 Eclipse。在初次启动时，需要设置工作空间，比如将工作空间设置为 D:\JSP workspace，如图 2.6 所示。

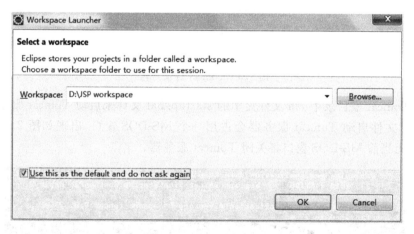

图 2.6　设置工作空间

2）配置 Eclipse

① 配置 Tomcat。启动 Eclipse，选择 Window/Preferences 命令，在打开的窗口中选择 Server 目录下的 Runtime Environments 选项，如图 2.7 所示。

② 单击 Add 按钮后，打开如图 2.8 所示的 New Server Runtime Environment 窗口，在此可以配置各种版本的 Web 服务器。

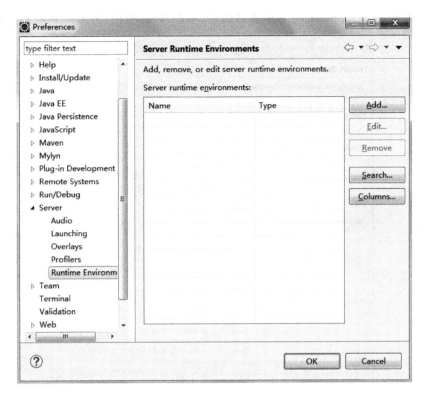

图 2.7　Preferences 窗口

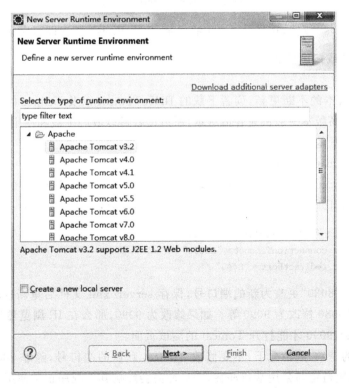

图 2.8　New Server Runtime Environment 窗口

③ 选择 Apache Tomcat v7.0 服务器版本，单击 Next 按钮，进入如图 2.9 所示界面。

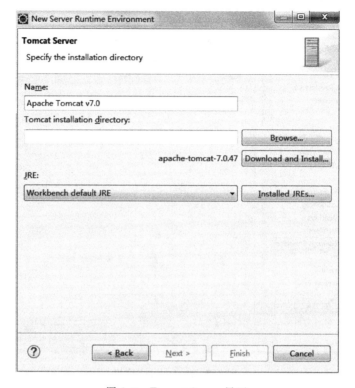

图 2.9　Tomcat Server 界面

④ 单击 Browse 按钮，选择 Tomcat 的目录，单击 Finish 按钮即可完成 Tomcat 配置。

4. 任务小结或知识扩展

1）软件版本

由于 Java 版本的不断更新，读者下载的 JDK、Tomcat 以及 Eclipse 的版本可能和本书使用的不同。但高版本一般都兼容低版本，所以读者可以放心下载和使用最新的版本，这些版本的安装和配置基本一致。

2）修改 Tomcat 的默认端口

8080 是 Tomcat 服务器默认占用的端口。但可以通过修改 Tomcat 的配置文件修改端口号。用记事本打开 conf 文件夹下的 server.xml 文件，找到以下代码：

```
<Connector port = "8080" protocol = "HTTP/1.1"
          connectionTimeout = "20000"
          redirectPort = "8443" />
```

将其中的 port="8080" 更改为新的端口号，保存 server.xml 文件后重新启动 Tomcat 服务器即可，比如将 8080 修改为 9090 等。如果修改为 9090，那么在 IE 浏览器地址栏中要输入 http://localhost:9090 才能打开 Tomcat 的测试页面。

需要说明的是，一般情况下，不要修改 Tomcat 默认的端口号，除非 8080 已经被占用。在修改端口时，应避免与公用端口冲突，一旦冲突会影响其他程序正常使用。

2.1.4 实践环节

修改 Tomcat 的端口号并测试。使用记事本打开 Tomcat 目录下的 conf 文件夹下的 server.xml 文件，找到以下代码：

```
<Connector port="8080" protocol="HTTP/1.1"
           connectionTimeout="20000"
           redirectPort="8443" />
```

将其中的 port="8080" 更改为新的端口号 9999，保存 server.xml 文件后重新启动 Tomcat 服务器。然后在 IE 浏览器地址栏中输入 http://localhost:9999 打开 Tomcat 的测试页面。

2.2 使用 Eclipse 开发 Web 应用

2.2.1 核心知识

1. JSP 文件

一个 JSP 文件中可以有普通的 HTML 标记、JSP 规定的标记以及 Java 程序。JSP 文件的扩展名是 .jsp，文件的名字必须符合标识符规定，即名字可以由字母、下划线、美元符号和数字组成。

2. JSP 运行原理

当 Web 服务器上的一个 JSP 页面第一次被客户端请求执行时，Web 服务器上的 JSP 引擎首先将 JSP 文件转译成一个 Java 文件，并将 Java 文件编译成字节码文件，然后执行字节码文件响应客户端的请求。当这个 JSP 页面再次被请求时，JSP 引擎将直接执行字节码文件响应客户端的请求，这也是 JSP 比 ASP 速度快的原因之一。

JSP 引擎以如下方式处理 JSP 页面。

（1）将 JSP 页面中的静态元素（HTML 标记）直接交给客户端浏览器执行显示。

（2）对 JSP 页面中的动态元素（Java 程序和 JSP 标记）进行必要的处理，将需要显示的结果发送给客户端浏览器。

2.2.2 能力目标

（1）使用 Eclipse 创建 Web 项目。
（2）在项目中创建 JSP 文件。
（3）发布项目到 Tomcat 服务器并运行。

2.2.3 任务驱动

任务的主要内容如下。
（1）创建项目。
（2）创建 JSP 文件。
（3）发布项目到 Tomcat 并运行。

1. 创建项目

(1) 启动 Eclipse,进入 Eclipse 开发界面。

(2) 选择主菜单中的 File/New/Project 命令,打开 New Project 窗口,在该窗口中选择 Web 节点下的 Dynamic Web Project 子节点,如图 2.10 所示。

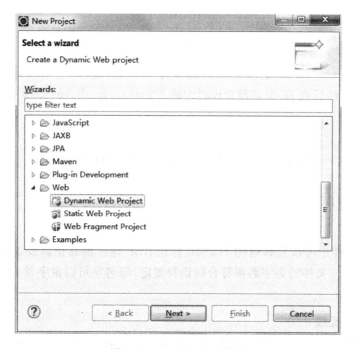

图 2.10 New Project 窗口

(3) 单击 Next 按钮,打开 New Dynamic Web Project 窗口,在该窗口的 Project name 文本框中输入项目名称,这里为 firstProject。选择 Target runtime 下拉列表框中的服务器,如图 2.11 所示。

(4) 单击 Finish 按钮,完成项目 firstProject 的创建。此时在 Eclipse 平台左侧的 Project Explorer 选项卡中,将显示项目 firstProject,依次展开各节点,可显示如图 2.12 所示的目录结构。

2. 创建 JSP 文件

firstProject 项目创建完成后,可以根据实际需要创建类文件、JSP 文件或者其他文件。这些文件的创建会在需要时介绍,下面将创建一个名为 myFirst.jsp 的 JSP 文件。

(1) 选中 firstProject 项目的 WebContent 节点,右击,在打开的快捷菜单中,选择 New/JSP File 命令,打开 New JSP File 窗口,在该窗口的 File name 文本框中输入文件名 myFirst.jsp,其他采用默认设置,单击 Finish 按钮完成 JSP 文件的创建。

(2) JSP 创建完成后,在 firstProject 项目的 WebContent 节点下,自动添加一个名为 myFirst.jsp 的 JSP 文件,同时,Eclipse 会自动将 JSP 文件在右侧的编辑框中打开。

(3) 将 myFirst.jsp 文件中的默认代码修改如下:

<% @ page language = "java" contentType = "text/html; charset = GBK" pageEncoding = "GBK" %>

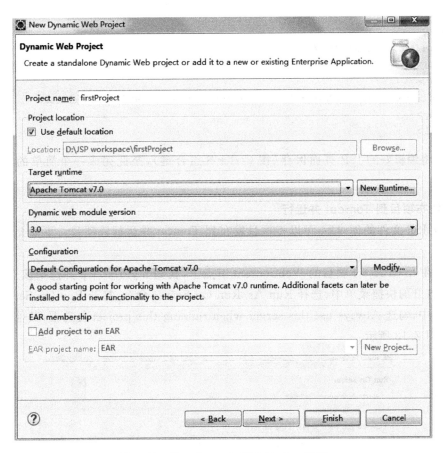

图 2.11　New Dynamic Web Project 窗口

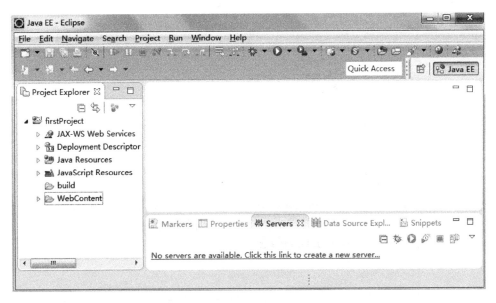

图 2.12　项目 firstProject 的目录结构

```
<html>
<head>
    <title>myFirst.JSP</title>
</head>
<body>
    <center>真高兴,忙乎半天了,终于要看到人生中第一个JSP页面了.</center>
</body>
</html>
```

(4) 将编辑好的 JSP 页面保存(按 Ctrl+S 组合键),至此完成一个简单的 JSP 程序创建。

3. 发布项目到 Tomcat 并运行

完成 JSP 文件的创建后,可以将项目发布到 Tomcat 并运行该项目。下面介绍具体的方法。

(1) 在 firstProject 项目的 WebContent 节点下,找到 myFirst.jsp 并选中该 JSP 文件,右击,在打开的快捷菜单中,选择 Run As/Run On Server 命令,打开 Run On Server 窗口,在该窗口中,勾选 Always use this server when running this project 复选框,其他采用默认设置,如图 2.13 所示。

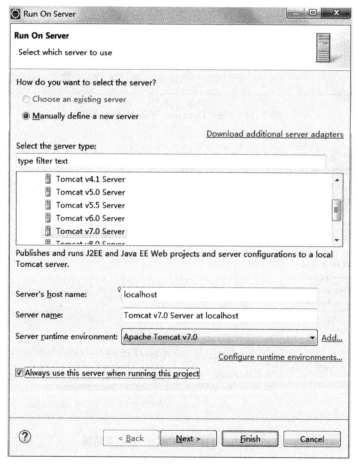

图 2.13　Run On Server 窗口

(2) 单击 Finish 按钮,即可通过 Tomcat 运行该项目,运行后的效果如图 2.14 所示。如果想在浏览器中运行该项目,可以将图 2.14 中的 URL 地址复制到浏览器的地址栏中,并按 Enter 键运行即可。

图 2.14 运行 firstProject 项目

注意:在 Eclipse 中,默认会将 Web 项目发布到 Eclipse 的工作空间的.metadata\.plugins\org.eclipse.wst.server.core\tmp0(或者是 tmp1)\wtpwebapps\ 目录下。

4. 任务小结或知识扩展

JSP 文件的默认编码格式为 ISO-8859-1,为了让页面支持中文,还需要将编码格式修改为 GBK 或 GB2312 或 UTF-8。

在一个项目的 WebContent 节点下可以创建多个 JSP 文件,另外 JSP 文件中使用到的图片文件、CSS 文件(层叠样式表)以及 JavaScript 文件都放在 WebContent 节点下。

2.2.4 实践环节

(1) 参照本节的任务内容 1,创建一个名为 sencondProject 的项目。

(2) 参照本节的任务内容 2,在 sencondProject 项目中创建一个名为 yourFirst.jsp 文件,在 JSP 页面中显示"不错!不错!自己能创建 JSP 文件了,并且可以发布运行了"。

(3) 参照本节的任务内容 3,发布并运行 sencondProject 项目。

2.3 小　　结

本章主要介绍了 JSP 集成开发环境的构建,用到的工具有 JDK、Tomcat 和 Eclipse。它们的安装顺序由先到后依次是 JDK、Tomcat、Eclipse。

Tomcat 是一个支持 JSP 的 Web 服务软件,该软件也称作 JSP 引擎或 JSP 容器。JSP 引擎是支持 JSP 程序的 Web 容器,负责运行 JSP 程序,并将有关结果发送给客户端。目前流行的 JSP 引擎有 Tomcat、Resin、JRun、WebSphere、WebLogic 等,本书使用的是 Tomcat 服务器。

习 题 2

1. 安装 Tomcat 服务器所在的计算机需要事先安装 JDK 吗?
2. Tomcat 服务器的默认端口号是什么?如果想修改该端口号,应该修改哪个文件?
3. First.jsp 和 first.jsp 是否是相同的 JSP 文件名字?
4. JSP 引擎是怎样处理 JSP 页面中 HTML 标记的?

第 3 章

JSP 语法

主要内容

(1) JSP 页面的基本构成。
(2) JSP 脚本元素。
(3) JSP 指令标记。
(4) JSP 动作标记。

一个 JSP 页面通常由 HTML 标记、JSP 注释、Java 脚本元素以及 JSP 标记 4 种基本元素组成。这 4 种基本元素在 JSP 页面中是如何被使用的,这将是本章介绍的重点。

本章涉及的 JSP 页面保存在工程 ch3 的 WebContent 中。

3.1 JSP 页面的基本构成

3.1.1 核心知识

在 HTML 静态页面文件中加入和 Java 相关的动态元素,就构成了一个 JSP 页面。一个 JSP 页面通常由以下 4 种基本元素组成。

(1) 普通的 HTML 标记。
(2) JSP 注释。
(3) Java 脚本元素,包括声明、Java 程序片和 Java 表达式。
(4) JSP 标记,如指令标记、动作标记和自定义标记等。

3.1.2 能力目标

能够识别 JSP 页面的基本元素。

3.1.3 任务驱动

1. 任务的主要内容

根据 task3_1.jsp 代码中的注释,识别 JSP 页面的基本元素。

2. 任务的代码模板

task3_1.jsp 的代码如下:

```jsp
<%@ page language = "java" contentType = "text/html; charset = GBK" pageEncoding = "GBK" %>
<!-- JSP 指令标记 -->
<jsp:include page = "a.jsp"/><!-- JSP 动作标记 -->
<%!
int i = 0;                          //数据声明
int add(int x, int y){              //方法声明
    return x + y;
}
%>
<html>  <!-- html 标记 -->
<head>
    <title>task3_1.jsp</title>
</head>
<body>
    <%
    i++;                            //Java 程序片
    int result = add(1,2);
    %>
    i 的值为<% = i %>   <%-- Java 表达式 --%>
    <br>
    1 + 2 的和为<% = result %>
</body>
</html>
```

a.jsp 的代码如下：

```jsp
<%@ page language = "java" contentType = "text/html; charset = GBK" pageEncoding = "GBK" %>
<html>
<head>
    <title>a.jsp</title>
</head>
<body>
    被 task3_1.jsp 动态引用
</body>
</html>
```

3. 任务小结或知识扩展

在 task3_1.jsp 代码中，看到许多 JSP 注释。注释能够增强 JSP 文件的可读性，便于 Web 项目的更新和维护。JSP 页面中常见的注释有以下两种。

1）HTML 注释

格式：

`<!-- HTML 注释 -->`

在标记符"<!--"和"-->"之间加入注释内容，就构成了 HTML 注释。

JSP 引擎对于 HTML 注释也要进行处理，也就是说不将它看作注释，如果其中有 JSP 代码，也将被 JSP 引擎处理。JSP 引擎把处理之后的 HTML 注释交给客户端，客户端通过浏览器查看 JSP 的源文件时，能够看到 HTML 注释。

2) JSP 注释

格式：

```
<%-- JSP 注释 --%>
```

在标记符"<%–"和"–%>"之间加入注释内容,就构成了 JSP 注释。

JSP 引擎将 JSP 注释当作真正的注释,在编译 JSP 页面时忽略这部分代码。因此,客户端通过浏览器查看 JSP 的源文件时,无法看到 JSP 注释。

3.1.4 实践环节

识别出如下 JSP 页面的基本元素：

```jsp
<%@ page language="java" contentType="text/html; charset=GBK" pageEncoding="GBK"%>
<!-- 学习 JSP 页面的基本构成 -->
<%!
    String content = "JSP 页面基本构成：";
%>
<html>
  <head>
    <title>practice3_1.jsp</title>
  </head>
  <body>
    <%
        content = content + "HTML 标记、JSP 注释、JSP 标记以及 Java 脚本元素";
    %>
    <%=content%>
  </body>
</html>
```

3.2 Java 程序片

3.2.1 核心知识

在标记符"<%"和"%>"之间插入的 Java 代码被称作 JSP 页面的 Java 程序片。Java 程序片格式如下：

```
<% Java 代码 %>
```

一个 JSP 页面可以有许多程序片,这些程序片将被 JSP 引擎(本书指 Tomcat 服务器)按顺序执行。在一个程序片中声明的变量称为 JSP 页面的局部变量,它们在 JSP 页面后继的所有程序片部分以及表达式部分都有效。

当多个客户请求一个 JSP 页面时,JSP 引擎为每个客户启动一个线程,不同的线程会分别执行该 JSP 页面中的 Java 程序片,程序片中的局部变量会在不同的线程中被分配不同的内存空间。因此,一个客户对 JSP 页面局部变量操作的结果,不会影响其他客户。Java 程序片执行原理如图 3.1 所示。

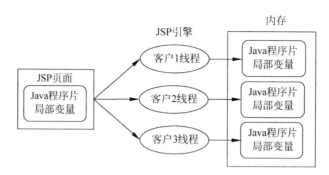

图 3.1　Java 程序片执行原理

3.2.2　能力目标

理解 Java 程序片的执行原理,掌握在 JSP 页面中如何使用 Java 程序片。

3.2.3　任务驱动

1. 任务的主要内容

编写一个 JSP 页面 task3_2.jsp,页面中存在一段 Java 程序片,该程序片内声明了一个整型的局部变量 x,初始值为 0。

2. 任务的代码模板

task3_2.jsp 的代码模板如下:

```
<%@ page language = "java" contentType = "text/html; charset = GBK" pageEncoding = "GBK" %>
<html>
<head>
    <title>task3_2.jsp</title>
</head>
    <body>
        【代码1】                          //Java 程序片开始
            【代码2】                      //声明 int 型局部变量 x,初始值为 0
            x++;
            out.print("x = " + x);        //在浏览器中输出 x 的值
        【代码3】                          <!-- Java 程序片结束 -->
    </body>
</html>
```

3. 任务小结或知识扩展

如果有 5 个客户请求 task3_2.jsp 页面,JSP 引擎会启动 5 个线程,页面中的 Java 程序片在每个线程中均会被执行一次,共计执行 5 次;在内存中,局部变量 x 对应 5 处不同的存储空间,初始值都为 0,且都只执行了一次自加运算。所以,5 个不同的客户看到的页面效果是相同的,如图 3.2 所示。

图 3.2　task3_2.jsp 页面的执行结果

有时可以根据需要将一个 Java 程序片分割成几个更小的程序片,以便在这些小的程序片之间再插入 JSP 页面的一些其他标记元素。例如,在浏览器中输出大小为 15×10 表格的代码如下:

```jsp
<%@ page language = "java" contentType = "text/html; charset = GBK" pageEncoding = "GBK" %>
<html>
    <head>
        <title>程序片分割</title>
    </head>
    <body>
        <table border = "1">
        <%
            for (int i = 1; i <= 10; i++)
            {
        %>
            <tr>
                <%
                    for (int j = 1; j <= 15; j++)
                    {
                        int temp = i * j;
                %>
                <td>
                <%
                    out.print(i * j);
                %>
                </td>
                <%
                    }
                %>
            </tr>
        <%
            }
        %>
        </table>
    </body>
</html>
```

4. 代码模板的参考答案

【代码 1】: `<%`
【代码 2】: `int x = 0;`
【代码 3】: `%>`

3.2.4 实践环节

编写一个 JSP 页面,在 JSP 页面中使用 Java 程序片输出 26 个小写的英文字母表。页面运行效果如图 3.3 所示。

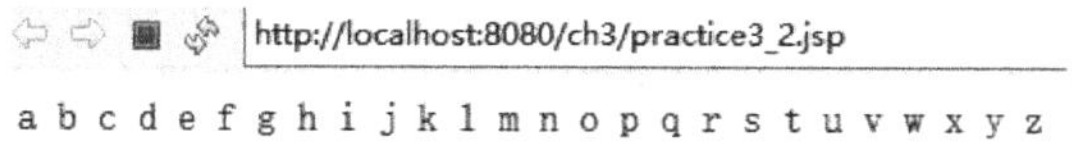

```
http://localhost:8080/ch3/practice3_2.jsp
a b c d e f g h i j k l m n o p q r s t u v w x y z
```

图 3.3 小写英文字母表

3.3 成员变量和方法的声明

在"<%!"和"%>"标记之间可以声明 JSP 的成员变量和方法。

3.3.1 核心知识

成员变量和方法的声明格式如下：

```
<%! 变量或方法定义 %>
```

在标记符"<%!"和"%>"之间声明的变量被称作 JSP 页面的成员变量，它们可以是 Java 语言允许的任何数据类型，例如：

```
<%!
    int n = 0;
    Date date;
%>
```

成员变量在整个 JSP 页面内都有效（与书写位置无关），因为 JSP 引擎将 JSP 页面转译成 Java 文件时，将这些变量作为类的成员变量，这些变量的内存空间直到服务器关闭才释放。因此，多个用户共享 JSP 页面的成员变量。任何用户对 JSP 页面成员变量操作的结果，都会影响其他用户。

在标记符"<%!"和"%>"之间声明的方法被称作 JSP 页面的成员方法，该方法在整个 JSP 页面内有效，但该方法内定义的变量仅在该方法内有效。

3.3.2 能力目标

理解 JSP 成员变量和方法的执行原理，学会使用 JSP 成员变量和方法。

3.3.3 任务驱动

1. 任务的主要内容

编写一个 JSP 页面 task3_3.jsp，页面中声明一个成员变量 n（初始值为 0）和方法 add（求两个整数的和）。另外，页面中还有一段 Java 程序片，在程序片声明一个局部变量 m，并且对成员变量 n 和局部变量 m 分别进行自加。

2. 任务的代码模板

task3_3.jsp 的代码模板如下：

```
<%@ page language="java" contentType="text/html; charset=GBK" pageEncoding="GBK" %>
<html>
    <head>
```

```
        <title>task3_3.jsp</title>
    </head>
    【代码1】                          //JSP声明开始
        【代码2】                      //声明成员变量n,初始值为0
        int add(int x,int y){
            return x+y;
        }
    【代码3】      <!-- JSP声明结束 -->
    <body>
        <%
            【代码4】                  //声明局部变量m,初始值为0
            n++;
            m++;
            int result = add(1,2);
            out.print("成员变量n的值为: "+n+"<br>");
            out.print("局部变量m的值为: "+m+"<br>");
            out.print("1+2 = "+result+"<br>"+"<br>");
            out.print("第"+n+"个客户");
        %>
    </body>
</html>
```

3. 任务小结或知识扩展

在 task3_3.jsp 中,变量 n 在标记符"<％!"和"％>"之间声明,因此是成员变量,被所有客户共享;变量 m 在标记符"<％"和"％>"之间声明,因此是局部变量,被每个客户独享。如果有 3 个客户请求这个 JSP 页面,则看到的效果如图 3.4 所示。

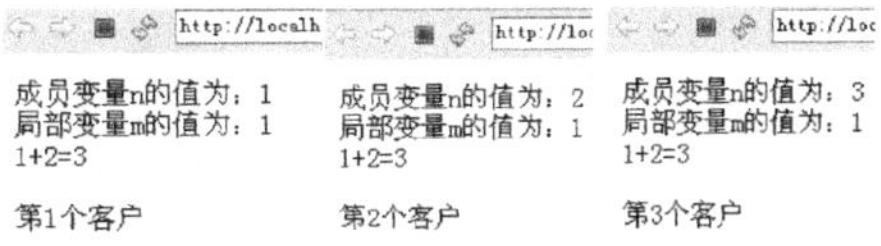

图 3.4　3 个客户请求 task3_3.jsp 页面的效果

从任务中可以看出 Java 程序片具有如下特点。
(1) 调用 JSP 页面声明的方法。
(2) 操作 JSP 页面声明的成员变量。
(3) 声明局部变量。
(4) 操作局部变量。

4. 代码模板的参考答案

【代码1】: <％!
【代码2】: int n = 0;
【代码3】: ％>
【代码4】: int m = 0;

3.3.4　实践环节

利用成员变量被所有客户共享这一性质,实现一个简单的计数器,页面效果如图 3.5

所示。

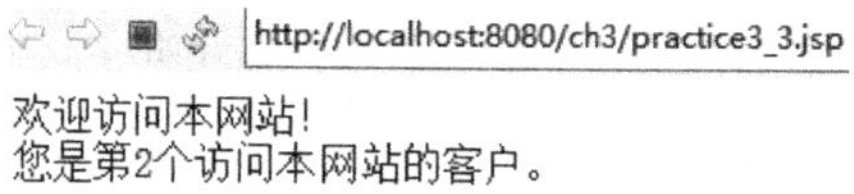

欢迎访问本网站！
您是第2个访问本网站的客户。

图 3.5　简单的计数器

3.4　Java 表达式

3.4.1　核心知识

在标记符"<%="和"%>"之间可以插入一个表达式，这个表达式必须能求值。表达式的值由 Web 服务器负责计算，并将计算结果用字符串形式发送到客户端，作为 HTML 页面的内容显示。

3.4.2　能力目标

能够灵活使用 Java 表达式计算数据并显示数据信息。

3.4.3　任务驱动

1. 任务的主要内容

在 task3_4.jsp 页面中使用 Java 表达式计算数据并显示数据信息，页面效果如图 3.6 所示。

2. 任务的代码模板

task3_4.jsp 的代码模板如下：

1+2等于3
若半径为5，则圆的面积是：78.5
2大于5是否成立：false

图 3.6　task3_4.jsp 的执行结果

```
<%@ page language="java" contentType="text/html; charset=GBK" pageEncoding="GBK"%>
<html>
    <head>
        <title>task3_4.jsp</title>
    </head>
    <%!
        int add(int x, int y){
            return x + y;
        }
    %>
    <body>
        1+2等于【代码1】  <br>  <!-- 使用Java表达式调用add方法计算1和2的值 -->
        若半径为5,则圆的面积是：【代码2】  <br>  <!-- 使用Java表达式计算出圆面积 -->
        2大于5是否成立：【代码3】  <br>  <!-- 使用Java表达式判断1是否大于2 -->
    </body>
</html>
```

3. 任务小结或知识扩展

Java 表达式中可以有算术表达式、逻辑表达式或条件表达式等。但使用 Java 表达式时,应该注意以下两点。

(1) 不可以在"<%="和"%>"之间插入语句,即输入的内容末尾不能以分号结束。
(2) "<%="是一个完整的符号,"<%"和"="之间不能有空格。

4. 代码模板的参考答案

【代码1】: <% = add(1,2) %>
【代码2】: <% = 3.14 * 5 * 5 %>
【代码3】: <% = 1>2 %>

3.4.4 实践环节

使用 Java 表达式显示出系统的当前时间。页面效果如图 3.7 所示。

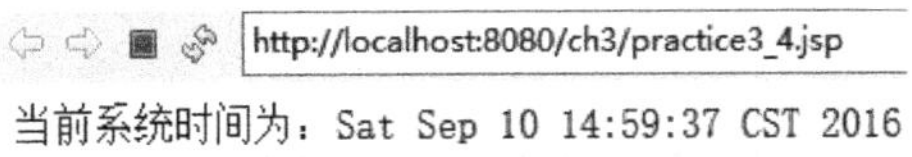

```
⇦ ⇨ ▪ 🗟  http://localhost:8080/ch3/practice3_4.jsp
当前系统时间为: Sat Sep 10 14:59:37 CST 2016
```

图 3.7　显示系统的当前时间

3.5　page 指令标记

page 指令标记用来定义整个 JSP 页面的一些属性和这些属性的值。可以用一个 page 指令指定多个属性的值,也可以使用多个 page 指令分别为每个属性指定值。page 指令的格式如下:

<% @ page 属性 1 = "属性 1 的值" 属性 2 = "属性 2 的值" …%>

或者

<% @ page 属性 1 = "属性 1 的值" %>
<% @ page 属性 2 = "属性 2 的值" %>
<% @ page 属性 3 = "属性 3 的值" %>
…
<% @ page 属性 n = "属性 n 的值" %>

page 指令的主要属性有 contentType、import、language 和 pageEncoding 等。

3.5.1 核心知识

page 指令标记通常定义的属性有以下 4 种。

1. 属性 contentType

JSP 页面使用 page 指令标记只能为 contentType 属性指定一个值,用来确定响应的 MIME 类型(MIME 类型就是设定某种文件用对应的一种应用程序来打开的方式类型)。当用户请求一个 JSP 页面时,服务器会告诉客户的浏览器使用 contentType 属性指定的 MIME 类型来解释执行所接收到的服务器为之响应信息。如果希望客户的浏览器使用 Word 应用程序打开用户请求的页面,就可以把 contentType 属性的值设置如下:

```
<%@page contentType = "application/msword;charset = GBK" %>
```

2. 属性 import

JSP 页面使用 page 指令标记可为 import 属性指定多个值，import 属性的作用是为 JSP 页面引入包中的类，以便在 JSP 页面的程序片、变量及方法声明或表达式中使用包中的类。

3. 属性 language

language 属性用来指定 JSP 页面使用的脚本语言，目前该属性的值只能取 java。

4. 属性 pageEncoding

contentType 中的 charset 是指网页内容从服务器发送给客户浏览器时用户所见到的该内容的编码；pageEncoding 是指 jsp 文件自身存储时所用的编码。

3.5.2 能力目标

读懂 page 指令标记为 JSP 页面指定的一些属性值。

3.5.3 任务驱动

1. 任务的主要内容

编写一个 JSP 页面 task3_5.jsp，当用户请求该页面时，客户浏览器启动本地的 PowerPoint 应用程序打开该页面。

2. 任务的代码模板

task3_5.jsp 的代码模板如下：

```
<%@ page
【代码1】= "java"
【代码2】= "application/vnd.ms-powerpoint; charset = GBK" pageEncoding = "GBK" %>
<!-- 代码1设置 language 属性,代码2设置 contentType 属性 -->
<%@ page【代码3】= "java.util.*" %>
<%@ page【代码4】= "java.io.*" %><!-- 代码3、4设置 import 属性 -->
<html>
    <head>
        <title>task3_5.jsp</title>
    </head>
    <body>
        在学习 page 指令标记时,请牢牢记住只能为 JSP 页面设置一个 contentType 属性值,可为 import 属性设置多个值
    </body>
</html>
```

3. 任务小结或知识扩展

使用 page 指令为 contentType 属性指定 MIME 类型，常见的有 text/html(HTML 解析器，所谓的网页形式)、text/plain(普通文本)、application/pdf(PDF 文档)、application/msword(Word 应用程序)、image/jpeg(JPEG 图形)、image/png(PNG 图像)、image/gif(GIF 图形)以及 application/vnd.ms-powerpoint(PowerPoint 应用程序)。

在 JSP 标准语法中，如果 pageEncoding 属性存在，那么 JSP 页面的字符编码方式就由 pageEncoding 决定，否则就由 contentType 属性的 charset 决定，如果 charset 也不存在，JSP 页面的字符编码方式就采用默认的 ISO-8859-1。

4. 代码模板的参考答案

【代码 1】：language
【代码 2】：contentType
【代码 3】：import
【代码 4】：import

3.5.4 实践环节

把任务中 task3_5.jsp 页面的 contentType 属性值指定为 application/msword，运行修改后的页面，并仔细观察运行结果。

3.6 include 指令标记

3.6.1 核心知识

一个网站中多个 JSP 页面有时需要显示同样信息，比如网站 Logo 或导航条等，为了便于网站维护，通常在这些 JSP 页面的适当位置嵌入一个内容相同的文件。include 指令标记的作用就是把 JSP 文件、HTML 网页文件或其他文本文件静态嵌入当前 JSP 网页中，该指令的语法格式如下：

```
<%@include file="文件的URL"%>
```

所谓静态嵌入就是"先包含后处理"，在编译阶段完成对文件嵌入，即先将当前 JSP 页面与被嵌入文件合并成一个新 JSP 页面，然后再由 JSP 引擎将新页面转化为 Java 文件处理并运行。

3.6.2 能力目标

理解静态嵌入的概念，并能够使用 include 指令标记在 JSP 网页中静态嵌入文件。

3.6.3 任务驱动

1. 任务的主要内容

编写两个 JSP 页面 task3_6.jsp 和 task3_6_1.jsp，在 task3_6.jsp 页面中使用 include 指令标记静态嵌入 task3_6_1.jsp 页面，访问 task3_6.jsp 页面，运行效果如图 3.8 所示。

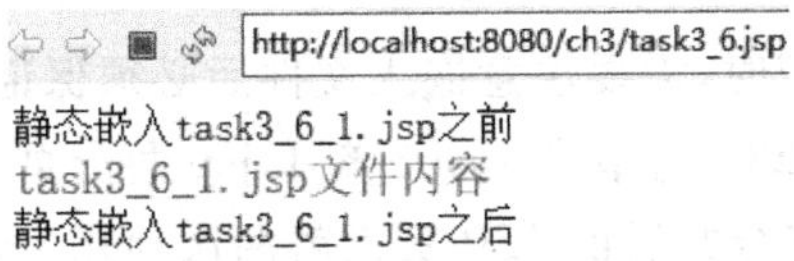

http://localhost:8080/ch3/task3_6.jsp

静态嵌入 task3_6_1.jsp 之前
task3_6_1.jsp 文件内容
静态嵌入 task3_6_1.jsp 之后

图 3.8　使用 include 指令标记

2. 任务的代码模板

task3_6.jsp 的代码模板如下：

```jsp
<%@ page language = "java" contentType = "text/html; charset = GBK" pageEncoding = "GBK" %>
<html>
    <head>
        <title>task3_6.jsp</title>
    </head>
    <body>
        静态嵌入 task3_6_1.jsp 之前
        <br>
        【代码1】 <!-- 使用 include 静态嵌入 task3_6_1.jsp -->
        <br>
        静态嵌入 task3_6_1.jsp 之后
    </body>
</html>
```

task3_6_1.jsp 的代码如下：

```jsp
<%@ page language = "java" contentType = "text/html; charset = GBK" pageEncoding = "GBK" %>
<html>
    <head>
        <title>task3_6_1.jsp</title>
    </head>
    <body>
        <font color = "red" size = 4>task3_6_1.jsp 文件内容</font>
    </body>
</html>
```

3. 任务小结或知识扩展

在该任务中，task3_6.jsp 页面静态嵌入 task3_6_1.jsp 页面，此时需要先将 task3_6_1.jsp 中所有代码全部嵌入 task3_6.jsp 的指定位置，形成一个新 JSP 文件，然后再将新文件提交给 JSP 引擎处理，如图 3.9 所示。

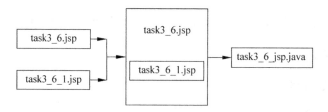

图 3.9 静态嵌入的原理

在使用 include 指令标记时，需要注意嵌入文件后必须保证新合成的 JSP 页面符合 JSP 语法规则，比如该任务中 task3_6.jsp 和 task3_6_1.jsp 两个页面的 page 指令就不能指定不同的 contentType 值，否则在合并后的 JSP 页面就两次使用 page 指令为 contentType 属性设置了不同属性值，导致语法错误。

4. 代码模板的参考答案

【代码1】：<%@ include file="task3_6_1.jsp" %>

3.6.4 实践环节

把任务中 task3_6.jsp 页面的 contentType 属性值修改为"application/msword;charset=GBK"，并运行修改后的页面。

3.7 include 动作标记

3.7.1 核心知识

动作标记 include 的作用是将 JSP 文件、HTML 网页文件或其他文本文件动态嵌入当前 JSP 网页中，该指令语法有以下两种格式：

<jsp:include page="文件的URL"/>

或

<jsp:include page="文件的URL">
    子标记
<jsp:include/>

关于子标记问题本书将在3.9节详细介绍，当动作标记 include 不需要子标记时，必须使用上述第一种形式。

所谓动态嵌入就是"先处理后包含"，在运行阶段完成对文件嵌入，即在把 JSP 页面转译成 Java 文件时，并不合并两个页面；而是在 Java 文件的字解码文件被加载并执行时，才去处理 include 动作标记中引入的文件。与静态嵌入方式相比，动态嵌入执行速度稍慢，但是灵活性较高。

3.7.2 能力目标

理解动态嵌入的概念，并能够使用 include 动作标记在 JSP 网页中动态嵌入文件。

3.7.3 任务驱动

1. 任务的主要内容

编写两个 JSP 页面 task3_7.jsp 和 task3_7_1.jsp，在 task3_7.jsp 页面中使用 include 动作标记动态嵌入 task3_7_1.jsp 页面。运行 task3_7.jsp 页面。

2. 任务的代码模板

task3_7.jsp 的代码模板如下：

<%@ page language="java" contentType="text/html; charset=GBK" pageEncoding="GBK" %>
<html>
    <head>

```
            <title>task3_7.jsp</title>
        </head>
        <body>
            动态嵌入 task3_7_1.jsp 之前
            <br>
            【代码1】      <!-- 使用 include 动作标记动态嵌入 task3_7_1.jsp -->
            <br>
            动态嵌入 task3_7_1.jsp 之后
        </body>
</html>
```

task3_7_1.jsp 的代码如下：

```
<%@ page language="java" contentType="text/html; charset=GBK" pageEncoding="GBK"%>
<html>
    <head>
        <title>task3_7_1.jsp</title>
    </head>
    <body>
        <font color="red" size=4>task3_7_1.jsp 文件内容</font>
    </body>
</html>
```

3. 任务小结或知识扩展

在该任务中，文件 task3_7.jsp 通过动作标记 include 动态嵌入了文件 task3_7_1.jsp，此时 JSP 引擎不会将两个文件合并成一个 JSP 页面，而是分别将文件 task3_7.jsp 和 task3_7_1.jsp 文件转化成对应的 Java 文件和字节码文件。当 JSP 解释器解释执行 task3_7.jsp 页面时，会遇到动作指令<jsp:include page="task3_7_1.jsp"/>对应的代码，此时才会执行 task3_7_1.jsp 页面对应的字节码文件，然后将执行的结果发送到客户端，并由客户端负责显示这些结果，所以 task3_7.jsp 和 task3_7_1.jsp 页面中 page 指令的 contentType 属性值可以不同。

4. 代码模板的参考答案

【代码1】：<jsp:include page="task3_7_1.jsp" />

3.7.4 实践环节

把任务中 task3_7.jsp 页面的 contentType 属性值修改为"application/msword; charset=GBK"，并运行修改后的页面。

3.8 forward 动作标记

3.8.1 核心知识

动作标记 forward 的作用是，从该标记出现处停止当前 JSP 页面继续执行，从而转向执行 forward 动作标记中 page 属性值指定的 JSP 页面。该标记有以下两种格式：

```
<jsp:forward page="文件的 URL"/>
```

或

```
<jsp:forward page="文件的 URL">
    子标记
<jsp:forward/>
```

当动作标记 forward 不需要子标记时，必须使用上述第一种形式。

3.8.2 能力目标

能够使用 forward 动作标记在 JSP 网页中实现页面跳转。

3.8.3 任务驱动

1. 任务的主要内容

编写 3 个 JSP 页面 task3_8.jsp、oddNumber.jsp 和 evenNumbers.jsp。在 task3_8.jsp 页面中使用 forward 标记转向 evenNumbers.jsp 或 oddNumber.jsp 页面，在 task3_8.jsp 页面中随机获取 0~10 的整数，如果该数为偶数就转向页面 evenNumbers.jsp，否则转向页面 oddNumber.jsp。首先访问 task3_8.jsp 页面。

2. 任务的代码模板

task3_8.jsp 的代码模板如下：

```
<%@ page language="java" contentType="text/html; charset=GBK" pageEncoding="GBK"%>
<html>
    <head>
        <title>task3_8.jsp</title>
    </head>
    <body>
        <%
            long i = Math.round(Math.random()*10);
            if(i%2==0){
                System.out.println("获得的整数是偶数,即将跳转到偶数页面 evenNumbers.jsp.");
        %>
            【代码1】    <!-- 使用 forward 标记转向 evenNumbers.jsp 页面 -->
        <%
            System.out.println("我是偶数尝试一下能看到我吗?");
            }
            else{
                System.out.println("获得的整数是奇数,即将跳转到奇数页面 oddNumber.jsp.");
        %>
            【代码2】    <!-- 使用 forward 标记转向 oddNumber.jsp 页面 -->
        <%
            System.out.println("我是奇数尝试一下能看到我吗?");
            }
        %>
    </body>
</html>
```

evenNumbers.jsp 的代码如下：

```jsp
<%@ page language = "java" contentType = "text/html; charset = GBK" pageEncoding = "GBK" %>
<html>
    <head>
        <title>evenNumbers.jsp</title>
    </head>
    <body>
        我是偶数页
    </body>
</html>
```

oddNumber.jsp 的代码如下：

```jsp
<%@ page language = "java" contentType = "text/html; charset = GBK" pageEncoding = "GBK" %>
<html>
    <head>
        <title>oddNumber.jsp</title>
    </head>
    <body>
        我是奇数页
    </body>
</html>
```

3. 任务小结或知识扩展

在该任务中，当用户请求查看页面 task3_8.jsp 时，如果获取的整数是偶数，那么只会在控制台上看到"获得的整数是偶数，即将跳转到偶数页面 evenNumbers.jsp。"这句话，当 JSP 引擎执行到＜jsp:forward page＝"evenNumbers.jsp.jsp" /＞语句时，会停止当前页面的执行，然后自动跳转到 evenNumbers.jsp 页面，并在客户端的浏览器上显示 evenNumbers.jsp 页面内容。如果获取的整数是奇数，那么只会在控制台上看到"获得的整数是奇数，即将跳转到奇数页面 oddNumber.jsp。"这句话，当 JSP 引擎执行到＜jsp:forward page＝"oddNumber.jsp" /＞语句时，会停止当前页面的执行，然后自动跳转到 oddNumber.jsp 页面，并在客户端的浏览器上显示 oddNumber.jsp 页面内容。

4. 代码模板的参考答案

【代码 1】：<jsp:forward page = "evenNumbers.jsp"/>
【代码 2】：<jsp:forward page = "oddNumber.jsp"/>

3.8.4 实践环节

将 task3_8.jsp 页面中的 forward 动作标记修改为 include 动作标记，并对修改前与修改后的运行结果进行比较。

3.9 param 动作标记

3.9.1 核心知识

动作标记 param 不能独立使用，但可以作为 include、forward 动作标记的子标记来使

用,该标记以"名字-值"对形式为对应页面传递参数。该标记格式如下:

```
<jsp:父标记 page = "接收参数页面的 URL">
    <jsp:param name = "参数名" value = "参数值"/>
<jsp:父标记/>
```

接收参数页面可以使用内置对象 request 调用 getParameter("参数名")方法获取动作标记 param 传递过来的参数值,内置对象将在本书第 4 章介绍。

3.9.2 能力目标

能够使用 param 动作标记作为 include、forward 动作标记的子标记为对应页面传递参数。

3.9.3 任务驱动

1. 任务的主要内容

编写两个 JSP 页面 task3_9.jsp 和 show.jsp,在 task3_9.jsp 页面中使用 include 动作标记动态包含文件 show.jsp,并向它传递一个名为 userName,值为 kazhafei 的参数;show.jsp 收到参数后,计算参数值的字符个数,并输出参数值和它的字符个数。运行 task3_9.jsp 页面,效果如图 3.10 所示。

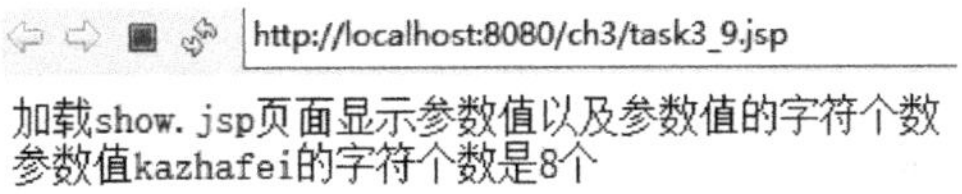

```
http://localhost:8080/ch3/task3_9.jsp
加载show.jsp页面显示参数值以及参数值的字符个数
参数值kazhafei的字符个数是8个
```

图 3.10 用 param 子标记向加载的文件传递值

2. 任务的代码模板

task3_9.jsp 的代码模板如下:

```jsp
<%@ page language = "java" contentType = "text/html; charset = GBK" pageEncoding = "GBK" %>
<html>
    <head>
        <title>task3_9.jsp</title>
    </head>
    <body>
        加载 show.jsp 页面显示参数值以及参数值的字符个数<br>
        <jsp:include page = "show.jsp">
            【代码 1】
        </jsp:include>
        <!-- 代码 1 使用 param 子标记传递一个名为 userName,值为 kazhafei 的参数 -->
    </body>
</html>
```

show.jsp 的代码模板如下:

```jsp
<%@ page language = "java" contentType = "text/html; charset = GBK" pageEncoding = "GBK" %>
<html>
    <head>
```

```
        <title>show.jsp</title>
    </head>
    <body>
        <%
            String name = request.getParameter("userName");
            int n = name.length();
            out.print("我是被加载的页面,负责计算参数值的长度<br>");
            out.print("参数值" + name + "的字符个数是" + n + "个");
        %>
    </body>
</html>
```

3. 任务小结或知识扩展

当使用 include 动作标记时经常会使用 param 子标记,以便向动态包含的 JSP 文件传递必要的参数值,这也体现出 include 动作标记比 include 指令标记灵活的特点。如果向页面传递多个参数,可以多次使用 param 子标记。格式如下:

```
<jsp:父标记 page = "接收参数页面的 URL">
    <jsp:param name = "参数名 1" value = "参数值 1"/>
    <jsp:param name = "参数名 2" value = "参数值 2"/>
    <jsp:param name = "参数名 3" value = "参数值 3"/>
    …
<jsp:父标记/>
```

4. 代码模板的参考答案

【代码 1】: <jsp:param value = "kazhafei" name = "userName"/>

3.9.4 实践环节

编写两个页面 practice3_9.jsp 和 computer.jsp,在页面 practice3_9.jsp 中使用 include 动作标记动态包含文件 computer.jsp,并向它传递一个矩形的长和宽;computer.jsp 收到参数后,计算矩形面积,并显示计算结果。运行 practice3_9.jsp 页面,效果如图 3.11 所示。

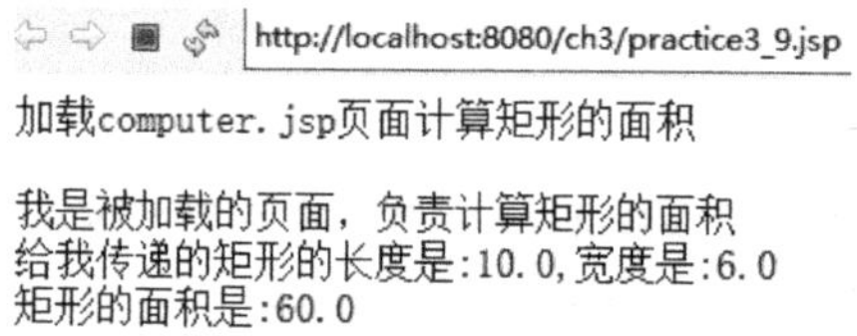

图 3.11 用 param 子标记向加载的文件传递多个值

3.10 小 结

本章主要介绍了 JSP 页面的组成、JSP 脚本元素和 JSP 标记。一个 JSP 页面通常由普通的 HTML 标记、JSP 注释、Java 脚本元素以及 JSP 标记组成。JSP 脚本元素包括 Java 程序片、JSP 页面成员变量与方法的声明、Java 表达式。JSP 标记包括指令标记和动作标记。

习 题 3

1. JSP 页面由哪几种主要元素组成？
2. 如果有 3 个用户访问一个 JSP 页面，则该页面中 Java 程序片将被执行几次？
3. "<%!"和"%>"之间声明的变量与"<%"和"%>"之间声明的变量有何不同？
4. 动作标记 include 和指令标记 include 的区别是什么？
5. 一个 JSP 页面中是否允许使用 page 指令为 contentType 属性设置多个值？是否允许使用 page 指令为 import 属性设置多个值？
6. 编写 3 个 JSP 页面：main.jsp、first.jsp 和 second.jsp，将 3 个 JSP 文件保存在同一个 Web 服务目录中，main.jsp 使用 include 动作标记加载 first.jsp 和 second.jsp 页面。first.jsp 页面可以画出一个表格，second.jsp 页面可以计算出两个正整数的最大公约数。当 first.jsp 被加载时获取 main.jsp 页面 include 动作标记的 param 子标记提供的表格的行数和列数，当 second.jsp 被加载时获取 main.jsp 页面 include 动作标记的 param 子标记提供的两个正整数值。

JSP 内置对象

主要内容

(1) 请求对象 request。
(2) 应答对象 response。
(3) 会话对象 session。
(4) 全局应用程序对象 application。

有些对象在 JSP 页面中不需要声明和实例化，可以直接在 Java 程序片和 Java 表达式部分使用，称这样的对象为 JSP 内置对象。JSP 内置对象由 Web 服务器负责实现和管理，JSP 自带了 9 个功能强大的内置对象，包括 request、response、session、application、out、page、pageContext、exception 和 config。本章主要学习前 4 种内置对象的使用方法。

本章涉及的 JSP 页面保存在工程 ch4 的 WebContent 中。

4.1 请求对象 request

request 内置对象是实现了 javax.servlet.ServletRequest 接口的一个实例。当用户请求一个 JSP 页面时，JSP 页面所在服务器将用户发出的所有请求信息封装在内置对象 request 中，使用该对象就可以获取用户提交的信息。

request 内置对象的常用方法如表 4.1 所示。

表 4.1 request 对象的常用方法

序号	方法	功能说明
1	Object getAttribute(String name)	返回指定属性的属性值
2	Enumeration getAtrributeNames( )	返回所有可用属性名的枚举
3	String getCharacterEncoding( )	返回字符编码方式
4	int getContentLength( )	返回请求体的字节数
5	String getContentType( )	返回请求体的 MIME 类型
6	ServletInputStream getInputStream( )	返回请求体中一行的二进制流
7	String getParameter(String name)	返回 name 指定参数的参数值

续表

序号	方 法	功 能 说 明
8	Enumeration getParameterNames()	返回可用参数名的枚举
9	String[] getParameterValues(String name)	返回包含参数 name 的所有值的数组
10	String getProtocol()	返回请求用的协议类型及版本号
11	String getServerName()	返回接受请求的服务器主机名
12	int getServerPort()	返回服务器接受此请求所用的端口号
13	String getRemoteAddr()	返回发送此请求的客户端 IP 地址
14	String getRemoteHost()	返回发送此请求的客户端主机名
15	void setAttribute(String key, Object obj)	设置属性的属性值
16	String getRealPath(String path)	返回一虚拟路径的真实路径

1. request 对象获取表单信息

1) String getParameter(String name)

该方法以字符串形式返回客户端传递的某个参数值,该参数名由 name 指定。

【**例 4.1**】 调用方法 getParameter(String name)获取表单信息。

例 4.1 页面文件 example4_1.jsp 的代码如下:

```
<%@ page language = "java" contentType = "text/html; charset = GBK" pageEncoding = "GBK" %>
<html>
    <head>
        <title>example4_1.jsp</title>
    </head>
    <body>
        <form action = "getValue.jsp">
            <input type = "text" name = "userName"/>
            <input type = "submit" value = "提交"/>
        </form>
    </body>
</html>
```

例 4.1 页面文件 getValue.jsp 的代码如下:

```
<%@ page language = "java" contentType = "text/html; charset = GBK" pageEncoding = "GBK" %>
<html>
    <head>
        <title>getValue.jsp</title>
    </head>
    <body>
        <%
            String name = request.getParameter("userName");
            out.println(name);
        %>
    </body>
</html>
```

2) String[]getParameterValues(String name)

该方法以字符串数组的形式返回客户端向服务器端传递的指定参数名的所有值。

【例 4.2】 调用方法 getParameterValues（String name）获取表单信息。

例 4.2 页面文件 example4_2.jsp 的代码如下：

```
<%@ page language = "java" contentType = "text/html; charset = GBK" pageEncoding = "GBK" %>
<html>
    <head>
        <title>example4_2.jsp</title>
    </head>
    <body>
        <form action = "getValues.jsp">
            选择您去过的城市：<br/>
            <input type = "checkbox" name = "cities" value = "beijing"/>北京
            <input type = "checkbox" name = "cities" value = "shanghai"/>上海
            <input type = "checkbox" name = "cities" value = "xianggang"/>香港
            <input type = "submit" value = "提交"/>
        </form>
    </body>
</html>
```

例 4.2 页面文件 getValues.jsp 的代码如下：

```
<%@ page language = "java" contentType = "text/html; charset = GBK" pageEncoding = "GBK" %>
<html>
    <head>
        <title>getValues.jsp</title>
    </head>
    <body>
        您去过的城市：<br>
        <%
            String yourCities[] = request.getParameterValues("cities");
            for(int i = 0; i < yourCities.length; i ++){
                out.println(yourCities[i] + "<br>");
            }
        %>
    </body>
</html>
```

2. NullPointerException 异常

如果不选择 example4_2.jsp 页面中的城市，直接单击"提交"按钮，那么 getValues.jsp 页面就会提示出现 NullPointerException 异常。为了避免在运行时出现 NullPointerException 异常，在 getValues.jsp 页面中使用如下代码：

```
if(yourCities != null){
    for(int i = 0; i < yourCities.length; i ++){
        out.print(yourCities[i] + "<br>");
    }
}
```

4.1.2 能力目标

能够灵活使用 request 内置对象获取客户提交的信息。

4.1.3 任务驱动

1. 任务的主要内容

编写两个 JSP 页面 task4_1.jsp 和 task4_1_1.jsp，task4_1.jsp 通过表单向 task4_1_1.jsp 提交输入的姓名和选择的城市，task4_1_1.jsp 负责获得表单提交的信息并显示。页面运行效果如图 4.1 所示。

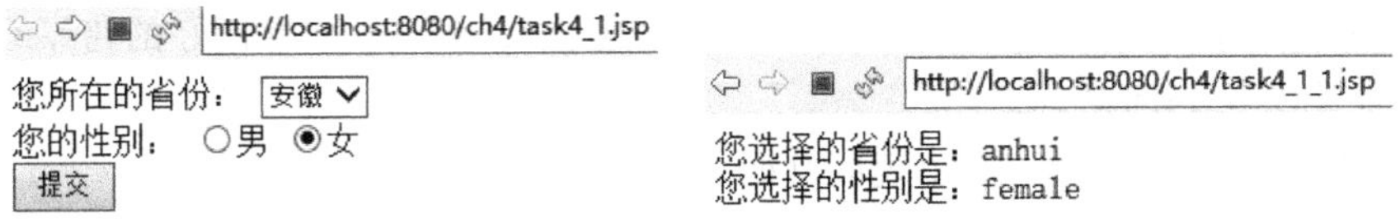

图 4.1 request 获得表单信息

2. 任务的代码模板

task4_1.jsp 的代码如下：

```
<%@ page language = "java" contentType = "text/html; charset = GBK" pageEncoding = "GBK" %>
<html>
    <head>
        <title>task4_1.jsp</title>
    </head>
    <body>
        <form action = "task4_1_1.jsp" method = "post">
            您所在的省份：
            <select name = "province">
                <option value = "liaoning">辽宁</option>
                <option value = "anhui">安徽</option>
                <option value = "shandong">山东</option>
            </select>
            <br>
            您的性别：
            <input type = "radio" name = "sex" value = "male">男
            <input type = "radio" name = "sex" value = "female">女
            <br>
            <input type = "submit" value = "提交">
        </form>
    </body>
</html>
```

task4_1_1.jsp 的代码如下：

```
<%@ page language = "java" contentType = "text/html; charset = GBK" pageEncoding = "GBK" %>
<html>
    <head>
        <title>task4_1_1.jsp</title>
    </head>
    <body>
        <%
            String myprovince =【代码 1】//获得 task4_1.jsp 页面选择的省份
```

```
        String mysex = 【代码2】//获得task4_1.jsp页面选择的性别
        %>
        您选择的省份是:<% = myprovince %><br>
        您选择的性别是:<% = mysex %>
    </body>
</html>
```

3. 任务小结或知识扩展

Java 的内核和 class 文件是基于 unicode 的,这使 Java 程序具有良好的跨平台性,但也产生了一些中文乱码问题。

1) 表单提交方式为 post 时出现的乱码

如果在 example4_1.jsp 页面的文本框中输入中文姓名,那么 getValue.jsp 页面获得的姓名是乱码。解决该乱码的常用方法有如下两种。

① 使用 setCharacterEncoding(String code)方法设置统一字符编码。request 对象提供了方法 setCharacterEncoding(String code)设置编码,其中参数 code 以字符串形式传入要设置的编码格式,但这种方法仅对于提交方式是 post 的表单(表单默认的提交方式是 get)有效。例如,使用该方法解决例 4.1 中的 getValue.jsp 页面出现的中文乱码问题,需要完成两项工作。

首先,将 example4_1.jsp 中的表单提交方式改为 post,具体代码如下:

```
<form action = "getValue.jsp" method = "post">
```

其次,在 getValue.jsp 中获取表单信息之前设置统一编码,具体代码如下:

```
request.setCharacterEncoding("GBK");
```

使用该方法解决中文乱码问题时,接受参数的每个页面都需要执行 request.setCharacterEncoding("GBK")。为了避免每个页面都编写 request.setCharacterEncoding("GBK")语句,可以使用过滤器对所有 JSP 页面进行编码处理。过滤器将在第 8 章讲解。

② 对获取的信息进行重新编码。通过内置对象 request 获取到字符串的值后,对该字符串使用 ISO-8859-1 重新编码,并把编码的结果存放到一个字节数组中,然后再将这个字节数组转化为字符串。例如,使用该方法解决例 4.1 中的 getValue.jsp 页面出现的中文乱码问题,具体代码如下:

```
String name = request.getParameter("userName");
byte b[] = name.getBytes("ISO - 8859 - 1");
name = new String(b);
```

2) 表单提交方式为 get 时出现的乱码

如果使用 get 方式提交中文,接收参数的页面也会产生乱码,这个乱码是 Tomcat 的内部编码格式 ISO-8859-1 导致的。Tomcat 会以 get 的默认编码方式 ISO-8859-1 对汉字进行编码,编码后追加到 URL 中,导致接收页面得到的参数值为乱码。解决方法如下。

① 对接收到的字符进行重新编码(post 提交方式解决办法的第二种)。

② 配置 Tomcat 下 server.xml 的 Connector 节点。在 Tomcat 的 server.xml 文件中找到 Connector 节点进行配置,配置前的 Connector 节点代码如下:

```
< Connector connectionTimeout = "20000" port = "8080" protocol = "HTTP/1.1" redirectPort =
"8443"/>
```

配置后的 Connector 节点代码如下:

```
< Connector URIEncoding = "GBK" connectionTimeout = "20000" port = "8080" protocol = "HTTP/1.1"
redirectPort = "8443"/>
```

3) JSP 页面通过 URL 传递中文参数的乱码问题

在项目中,可能会遇到在 JSP 页面跳转中传递中文字符。例如:

http://ch4/test1.jsp?act = test&type = 测试

解决 URL 传递参数乱码问题的方法如下:
① 配置 Tomcat 下 server.xml 的 Connector 节点(get 提交方式解决办法的第二种)。
② 对 URL 中的中文字符进行编码。

【例 4.3】 对 URL 中的中文字符进行编码。

例 4.3 传递参数页面的代码如下:

```
<% @ page language = "java" contentType = "text/html; charset = GBK" pageEncoding = "GBK" %>
<% @ page import = "java.net. * " %>
< html >
    < head >
        < title > URL 传参 </title >
    </head >
    < body >
        < a href = "test.jsp?a = <% = URLEncoder.encode("苹果") %>">URL 传参 </a>
    </body >
</html >
```

例 4.3 接收参数页面的代码如下:

```
<% @ page language = "java" contentType = "text/html; charset = GBK" pageEncoding = "GBK" %>
<% @ page import = "java.net. * " %>
< html >
    < head >
        < title > test.jsp </title >
    </head >
    < body >
        <%
        String a = URLDecoder.decode(request.getParameter("a"));
        out.println(new String(a.getBytes("ISO - 8859 - 1"), "GBK"));
        %>
    </body >
</html >
```

注意:上述中文乱码解决方法不要重复或交叉使用。例如,既配置 Tomcat 下 server.xml 的 Connector 节点,又对 URL 中的中文字符进行编码。

4) 字符集

为了更好地理解中文乱码的解决方法,需要了解几种常用的字符集。

① ASCII。ASCII(American Standard Code for Information Interchange,美国信息互换标准代码),是基于常用英文字符的一套编码。

② ISO-8859-1。ISO-8859-1 编码通常叫作 Latin-1,除收录 ASCII 字符外,还增加了其他一些语言和地区需要的字符。该编码是 Tomcat 服务器默认采用的字符编码。

③ GB2312。GB2312 码是中华人民共和国国家标准汉字信息交换用编码,简称国标码,是由国家标准总局发布的关于汉字的编码,通行于中国大陆和新加坡。

④ GBK。GBK 编码规范,除了完全兼容 GB2312,还对繁体中文和一些不常用的字符进行了编码。GBK 是现阶段 Windows 和其他一些中文操作系统的默认字符集。

⑤ Unicode。Unicode 为统一的字符编码标准集,为地球上几乎所有地区每种语言中的每个字符设定了统一并且唯一的编码,以满足跨语言、跨平台进行文本转换、处理的要求。

⑥ UTF-8。UTF-8 是 Unicode 的一种变长字符编码。用在网页上可以同一页面显示中文和其他语言。当处理包含多国文字的信息页面时一般选择用 UTF-8。

4. 代码模板的参考答案

【代码 1】: request.getParameter("province");
【代码 2】: request.getParameter("sex");

4.1.4 实践环节

使用两种方法(设置统一编码和重新编码)解决 example4_1.jsp 和 getValue.jsp 页面中出现的中文乱码问题。

4.2 应答对象 response

4.2.1 核心知识

当用户请求服务器一个页面时,会提交一个 HTTP 请求,服务器收到请求后,返回 HTTP 响应。request 对象对请求信息进行封装,与 request 对象对应的对象是 response 对象。response 对象对用户的请求做出动态响应。动态响应通常有动态改变 contentType 属性值、设置响应表头和 response 重定向。

1. 动态改变 contentType 属性值

JSP 页面用 page 指令标记设置了页面的 contentType 属性值,response 对象按照此属性值的方式对客户做出响应。在 page 指令中只能为 contentType 属性指定一个值。如果想动态改变 contentType 属性值,换一种方式来响应客户,可以让 response 对象调用 setContentType(String s)方法来重新设置 contentType 的属性值。

2. 设置响应表头(HTTP 文件头)

response 对象可以通过方法 setHeader(String name, String value)设置指定名字的 HTTP 文件头的值,以此来操作 HTTP 文件头。如果希望某页面每 3 秒刷新一次,那么在该页面中添加如下代码:

```
response.setHeader("refresh","3");
```

3. response 重定向

需要将用户引导至另一个页面时,可以使用 reponse 对象的 sendRedirect(String url) 方法实现用户的重定向。例如,用户输入的表单信息不完整或有误,应该再次被重定向到输入页面。

4.2.2 能力目标

能够灵活使用 response 内置对象动态响应用户的请求。

4.2.3 任务驱动

任务 1:动态改变 contentType 属性值

(1) 任务的主要内容。

编写一个 JSP 页面 task4_2_1.jsp,客户端通过单击页面上的不同按钮,可以改变页面响应的 MIME 类型。当单击 word 按钮时,JSP 页面动态地改变 contentType 的属性值为 application/msword,浏览器启用本地的 Word 软件来显示当前页面内容;当单击 excel 按钮时,JSP 页面动态地改变 contentType 的属性值为 application/vnd.ms-excel,浏览器启用本地的 Excel 软件来显示当前页面内容。效果如图 4.2 所示。

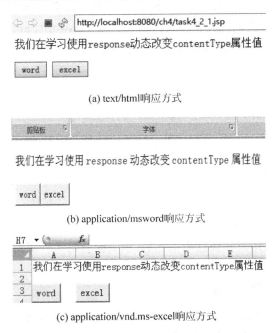

(a) text/html响应方式

(b) application/msword响应方式

(c) application/vnd.ms-excel响应方式

图 4.2 任务 task4_2_1 的效果图

(2) 任务的代码模板。

task4_2_1.jsp 的代码模板如下:

```
<%@ page language = "java" contentType = "text/html; charset = GBK" pageEncoding = "GBK" %>
<html>
    <head>
```

```
            <title>task4_2_1.jsp</title>
        </head>
        <body>
            <form action = "" method = "post">
                <p>我们在学习使用 response 动态改变 contentType 属性值
                <p>
                <input type = "submit" value = "word" name = "submit">
                <input type = "submit" value = "excel" name = "submit">
                <%
                    String str = request.getParameter("submit");
                    if ("word".equals(str)) {
                    【代码 1】//response 调用 setContentType 方法设置 MIME 类型为 application/msword
                    }else if ("excel".equals(str)) {
                    【代码 2】/* response 调用 setContentType 方法设置 MIME 类型为 application/ vnd.
                        ms - excel */
                    }
                %>
            </form>
        </body>
</html>
```

(3) 任务小结或知识扩展。

response 对象调用 setContentType(String s)方法来重新设置网页响应的 MIME 类型。常见的 MIME 类型有 text/html、application/msword、application/vnd. ms-excel、image/gif、image/jpeg、application/vnd. ms-powerpoint、application/x-shockwave-flash、application/pdf 等。

(4) 代码模板的参考答案。

【代码 1】: response.setContentType("application/msword");
【代码 2】: response.setContentType("application/vnd.ms - excel");

任务 2：设置响应表头

(1) 任务的主要内容。

编写一个 JSP 页面 task4_2_2.jsp，在该页面中使用 response 对象设置一个响应头 refresh，其值是 3。那么用户收到这个头之后，该页面会每 3 秒刷新一次。

(2) 任务的代码模板。

task4_2_2.jsp 的代码模板如下：

```
<%@ page language = "java" contentType = "text/html; charset = GBK" pageEncoding = "GBK" %>
<%@ page import = "java.util.*" %>
<html>
    <head>
        <title>task4_2_2.jsp</title>
    </head>
    <body>
        <h2>该页面每 3 秒刷新 1 次</h2>
        <p>现在的秒钟时间是:
```

```
    <%
        Date d = new Date();
        out.print("" + d.getSeconds());
    【代码1】           //使用response对象设置一个响应头"refresh",其值是"3"
    %>
    </body>
</html>
```

(3) 任务小结或知识扩展。

有时候希望从当前页面几秒后自动跳转到另一个页面。比如,打开one.jsp页面3秒后,自动跳转到another.jsp页面(one.jsp与another.jsp在同一个Web服务目录下)。这该如何实现呢？只需要为one.jsp设置一个响应头即可,也就是在one.jsp页面中添加如下代码：

response.setHeader("refresh","3;url = another.jsp");

(4) 代码模板的参考答案。

【代码1】: response.setHeader("refresh","3");

任务3：重定向

(1) 任务的主要内容。

编写两个JSP页面task4_2_3.jsp和enter.jsp,如果在页面task4_2_3.jsp中输入正确的密码2016manpinde,单击Come on按钮后提交给页面enter.jsp,如果输入不正确,重新定向到task4_2_3.jsp页面。先运行task4_2_3.jsp页面,页面效果如图4.3所示。

(a) task4_2_3.jsp页面

(b) enter.jsp页面

图4.3 页面效果图

(2) 任务的代码模板。

task4_2_3.jsp的代码如下：

```
<%@ page language = "java" contentType = "text/html; charset = GBK" pageEncoding = "GBK" %>
<html>
    <head>
        <title>task4_2_3.jsp</title>
    </head>
    <body>
        <form action = "enter.jsp" method = "post" name = form>
```

```
            <p>
                输入密钥:
            <br>
            <input type = "password" name = "pwd"/>
            <input type = "submit" value = "Come on">
        </form>
    </body>
</html>
```

enter.jsp 的代码模板如下:

```
<%@ page language = "java" contentType = "text/html; charset = GBK" pageEncoding = "GBK" %>
<html>
    <head>
        <title>enter.jsp</title>
    </head>
    <body>
        <%
            String str = request.getParameter("pwd");
            if (!"2016manpinde".equals(str)) {
                【代码1】        //重定向到 task4_2_3.jsp 页面重新输入密码
            } else {
                out.print("2016 年是蛮拼的一年!");
            }
        %>
    </body>
</html>
```

(3) 任务小结或知识扩展。

response 对象的 sendRedirect 方法是在用户的浏览器端工作的,Web 服务器要求浏览器重新发送一个到被定向页面的请求。在浏览器地址栏上会出现重定向页面的 URL,且为绝对路径。

forward 动作标记也可以实现页面的跳转,如<jsp:forward page = "info.jsp"/>。但使用 forward 动作标记与 response 对象调用 sendRedirect 不同。两者的区别如下。

① forward 为服务器端跳转,浏览器地址栏不变;sendRedirect 为客户端跳转,浏览器地址栏改变为新页面的 URL。

② 执行到 forward 标记出现处停止当前 JSP 页面的继续执行,而转向标记中 page 属性指定的页面;sendRedirect 是所有代码执行完毕之后再跳转。

③ 使用 forward,request 请求信息能够保留到下一个页面;使用 sendRedirect 不能保留 request 请求信息。

④ forward 传递参数的格式如下:

```
<jsp:forward page = "info.jsp">
    <jsp:param name = "no" value = "001"/>
    <jsp:param name = "age" value = "18"/>
</jsp:forward>
```

response 对象的 sendRedirect 传递参数的方式如下：

response.sendRedirect("info.jsp?sno = 001&sage = 18");

（4）代码模板的参考答案。

【代码1】：response.sendRedirect("task4_2_3.jsp");

4.2.4 实践环节

编写3个JSP页面login_1.jsp、server.jsp和loginSuccess.jsp。在页面login_1.jsp中输入用户名和密码，单击"提交"按钮将输入的信息提交给页面server.jsp。在server.jsp页面中进行登录验证：如果输入正确（用户名zhangsan，密码123），提示"成功登录，3秒钟后进入loginSuccess.jsp页面"，如果输入不正确，重新定向到login_1.jsp页面。先运行login_1.jsp页面，页面运行效果如图4.4所示。

图 4.4 页面效果图

4.3 会话对象 session

浏览器与Web服务器之间使用HTTP协议进行通信。HTTP是一种无状态协议，客户向服务器发出请求（request），服务器返回响应（response），连接就被关闭了，在服务器端不保留连接的相关信息。所以服务器必须采取某种手段来记录每个客户的连接信息。Web服务器可以使用内置对象session来存放有关连接的信息。session对象指的是客户端与服务器端的一次会话，从客户端连到服务器的一个Web应用程序开始，直到客户端与服务器断开为止。

4.3.1 核心知识

1. session 对象的 ID

Web服务器会给每一个用户自动创建一个session对象，为每个session对象分配一个唯一标识的String类型的session ID，这个ID用于区分其他用户。这样每个用户都对应着一个 session 对象，不同用户的 session 对象互不相同。session 对象调用 getId()方法就可以获取当前 session 对象的 ID。

【例 4.4】 编写 3 个 JSP 页面 example4_4_1.jsp、example4_4_2.jsp 和 example4_4_3.jsp，其中，example4_4_2.jsp 存放在目录 tom 中，example4_4_3.jsp 存放在目录 cat 中。用户首先访问 example4_4_1.jsp 页面，从该页面链接到 example4_4_2.jsp 页面，然后再从 example4_4_2.jsp 页面链接到 example4_4_3.jsp，效果如图 4.5 所示。

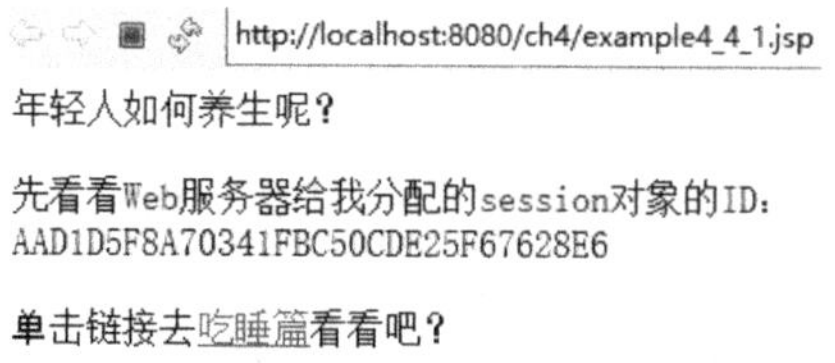
http://localhost:8080/ch4/example4_4_1.jsp
年轻人如何养生呢?
先看看Web服务器给我分配的session对象的ID:
AAD1D5F8A70341FBC50CDE25F67628E6
单击链接去吃睡篇看看吧?

(a) example4_4_1.jsp页面效果

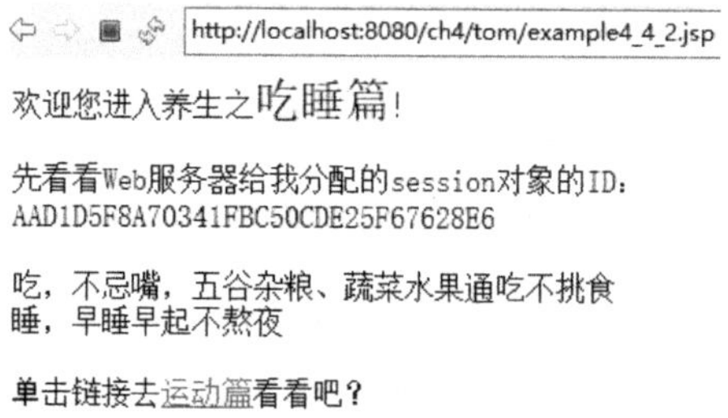
http://localhost:8080/ch4/tom/example4_4_2.jsp
欢迎您进入养生之吃睡篇!
先看看Web服务器给我分配的session对象的ID:
AAD1D5F8A70341FBC50CDE25F67628E6
吃，不忌嘴，五谷杂粮、蔬菜水果通吃不挑食
睡，早睡早起不熬夜
单击链接去运动篇看看吧?

(b) example4_4_2.jsp页面效果

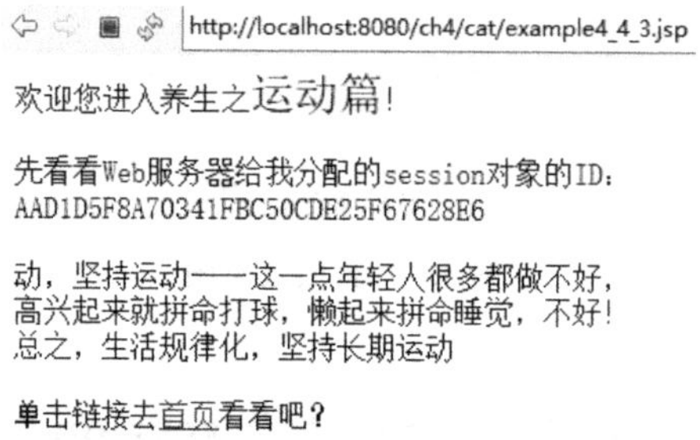
http://localhost:8080/ch4/cat/example4_4_3.jsp
欢迎您进入养生之运动篇!
先看看Web服务器给我分配的session对象的ID:
AAD1D5F8A70341FBC50CDE25F67628E6
动，坚持运动——这一点年轻人很多都做不好，
高兴起来就拼命打球，懒起来拼命睡觉，不好!
总之，生活规律化，坚持长期运动
单击链接去首页看看吧?

(c) example4_4_3.jsp页面效果

图 4.5 获取 session 对象的 ID

例 4.4 页面文件 example4_4_1.jsp 的代码如下：

```
<%@ page language = "java" contentType = "text/html; charset = GBK" pageEncoding = "GBK" %>
<html>
    <head>
        <title>example4_4_1.jsp</title>
    </head>
    <body>
        年轻人如何养生呢?<br><br>
        先看看 Web 服务器给我分配的 session 对象的 ID:
        <%
            String id = session.getId();
        %>
```

```
    <br>
    <% = id %>
    <br><br>
    单击链接去<a href = "tom/example4_4_2.jsp">吃睡篇</a>看看吧?
  </body>
</html>
```

例 4.4 页面文件 example4_4_2.jsp 的代码如下：

```
<%@ page language = "java" contentType = "text/html; charset = GBK" pageEncoding = "GBK" %>
<html>
  <head>
    <title>example4_4_2.jsp</title>
  </head>
  <body>
    欢迎您进入养生之<font size = 5>吃睡篇</font>!<br><br>
    先看看 Web 服务器给我分配的 session 对象的 ID:
    <%
        String id = session.getId();
    %>
    <br>
    <% = id %>
    <br><br>
    吃,不忌嘴,五谷杂粮、蔬菜水果通吃不挑食<br>
    睡,早睡早起不熬夜<br><br>
    点击链接去<a href = "../cat/example4_4_3.jsp">运动篇</a>看看吧?
  </body>
</html>
```

例 4.4 页面文件 example4_4_3.jsp 的代码如下：

```
<%@ page language = "java" contentType = "text/html; charset = GBK" pageEncoding = "GBK" %>
<html>
  <head>
    <title>example4_4_3.jsp</title>
  </head>
  <body>
    欢迎您进入养生之<font size = 5>运动篇</font>!<br><br>
    先看看 Web 服务器给我分配的 session 对象的 ID:
    <%
        String id = session.getId();
    %>
    <br>
    <% = id %>
    <br><br>
    动,坚持运动——这一点年轻人很多都做不好,<br>高兴起来就拼命打球,懒起来拼命睡觉,不好!<br>
    总之,生活规律化,坚持长期运动<br><br>
    单击链接去<a href = "../example4_4_1.jsp">首页</a>看看吧?
  </body>
</html>
```

从例 4.4 各个页面的运行结果可以看出，一个用户在同一个 Web 服务目录中只有一个 session 对象，当用户访问相同 Web 服务目录的其他页面时，Web 服务器不会再重新分配 session 对象，直到用户关闭浏览器或这个 session 对象达到了它的生存期限。当用户重新打开浏览器再访问该 Web 服务目录时，Web 服务器为该客户再创建一个新的 session 对象。

需要注意的是，同一用户在多个不同的 Web 服务目录中所对应的 session 对象是不同的，一个服务目录对应一个 session 对象。

2. session 对象存储数据

使用 session 对象可以保存用户在访问某个 Web 服务目录期间的有关数据。有关处理数据的方法如下。

1) public void setAttribute(String key, Object obj)

将参数 obj 指定的对象保存到 session 对象中，key 为所保存的对象指定一个关键字。若保存的两个对象关键字相同，则先保存的对象被清除。

2) public Object getAttibute(String key)

获取 session 中关键字是 key 的对象。

3) public void removeAttribute(String key)

从 session 中删除关键字 key 所对应的对象。

4) public Enumeration getAttributeNames()

产生一个枚举对象，该枚举对象可使用方法 nextElemets()遍历 session 中各个对象所对应的关键字。

【例 4.5】 使用 session 对象模拟在线考试系统。编写 3 个 JSP 页面 example4_5_1.jsp、example4_5_2.jsp 和 example4_5_3.jsp，在 example4_5_1.jsp 页面中考试，在 example4_5_2.jsp 页面中显示答题结果，在 example4_5_3.jsp 页面中计算并公布考试成绩。首先运行 example4_5_1.jsp 页面，效果如图 4.6 所示。

例 4.5 页面文件 example4_5_1.jsp 的代码如下：

```
<%@ page language="java" contentType="text/html; charset=GBK" pageEncoding="GBK"%>
<html>
    <head>
        <title>example4_5_1.jsp</title>
    </head>
    <body>
        <form action="example4_5_2.jsp" method="post">
            考号：
            <input type="text" name="id"/>
            <p>
            一、单项选择题(每题 2 分)
            <br/><br/>
            1.下列哪个方法是获取 session 中关键字是 key 的对象( ).
            <br/>
            <input type="radio" name="one" value="A"/>
            A.public void setAttribute(String key, Object obj)<br/>
            <input type="radio" name="one" value="B"/>
```

第 4 章 JSP 内置对象

(a) 试卷页面

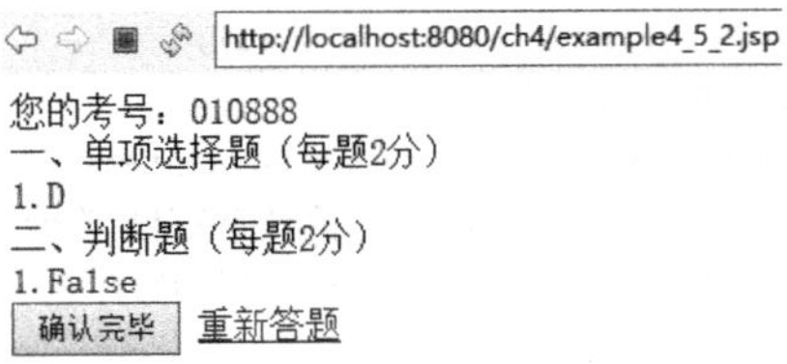

(b) 确认页面

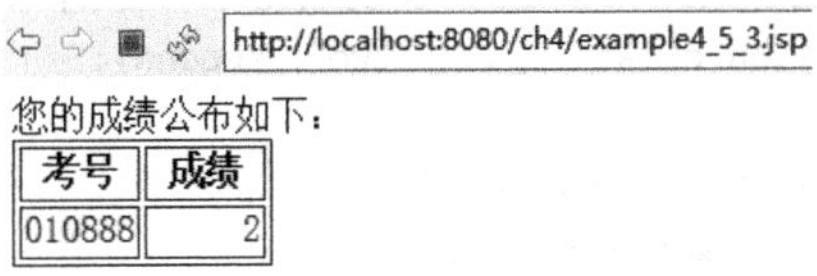

(c) 成绩公布页面

图 4.6　session 对象模拟考试系统

```
        B. public void removeAttribute(String key)<br/>
        <input type = "radio" name = "one" value = "C"/>
        C. public Enumeration getAttributeNames()<br/>
        <input type = "radio" name = "one" value = "D"/>
        D. public Object getAttibute(String key)<br/>
    </p>
    <p>
        二、判断题(每题2分)
        <br/><br/>
        1.同一客户在多个Web服务目录中,所对应的session对象是互不相同的。
        <br/>
        <input type = "radio" name = "two" value = "True"/>
        True
        <input type = "radio" name = "two" value = "False"/>
        False
    </p><br/>
    <input type = "submit" value = "提交" name = submit>
    <input type = "reset" value = "重置" name = reset>
</form>
</body>
```

</html>

例 4.5 页面文件 example4_5_2.jsp 的代码如下：

```jsp
<%@ page language="java" contentType="text/html; charset=GBK" pageEncoding="GBK"%>
<html>
    <head>
        <title>example4_5_2.jsp</title>
    </head>
    <body>
        <form action="example4_5_3.jsp" method="post">
            <%
                //考号
                String id = request.getParameter("id");
                //把考号 id 以"id"为关键字存储到 session 对象中
                session.setAttribute("id", id);
                //单项选择第一题
                String first = request.getParameter("one");
                //把答案 first 以"one"为关键字存储到 session 对象中
                session.setAttribute("one", first);
                //判断第一题
                String second = request.getParameter("two");
                //把答案 second 以"two"为关键字存储到 session 对象中
                session.setAttribute("two", second);
            %>
            您的考号：<%= id %><br/>
            一、单项选择题(每题 2 分)
            <br/>
            1.<%= first %>
            <br />
            二、判断题(每题 2 分)
            <br />
            1.<%= second %><br/>
            <input type="submit" value="确认完毕"/>
            <a href="example4_5_1.jsp">重新答题</a>
        </form>
    </body>
</html>
```

例 4.5 页面文件 example4_5_3.jsp 的代码如下：

```jsp
<%@ page language="java" contentType="text/html; charset=GBK" pageEncoding="GBK"%>
<html>
    <head>
        <title>example4_5_3.jsp</title>
    </head>
    <body>
        <%
            //获取考号
            //获取 session 中关键字是 id 的对象(考号)
            String id = (String) session.getAttribute("id");
            //计算成绩
```

```
            int sum = 0;
            //如果单项选择第一题选中 D 选项,得 2 分
            //获取 session 中关键字是 one 的对象(选择答案)
            String first = (String) session.getAttribute("one");
            if ("D".equals(first)) {
                sum += 2;
            }
            //如果判断第一题选中 True,得 2 分
            //获取 session 中关键字是 two 的对象(判断答案)
            String second = (String) session.getAttribute("two");
            if ("True".equals(second)) {
                sum += 2;
            }
        %>
        您的成绩公布如下:
        <table border = "1">
            <tr>
                <th width = "50%">
                    考号
                </th>
                <th width = "50%">
                    成绩
                </th>
            </tr>
            <tr>
                <td><% = id %></td>
                <td align = "right"><% = sum %></td>
            </tr>
        </table>
    </body>
</html>
```

3. session 对象的生存期限

一个用户在某个 Web 服务目录中的 session 对象的生存期限依赖于以下几个因素。

(1) 用户是否关闭浏览器。

(2) session 对象是否调用 invalidate()方法。

(3) session 对象是否达到设置的最长"发呆"时间。

与 session 对象生命周期相关的方法如表 4.2 所示。

表 4.2 session 对象的方法

序号	方法	功能说明
1	long getCreationTime( )	返回 session 创建时间
2	long getLastAccessedTime ( )	返回此 session 里客户端最近一次请求时间
3	int getMaxInactiveInterval( )	返回两次请求间隔时间(单位是秒)
4	void invalidate( )	使 session 失效
5	boolean isNew( )	判断客户端是否已经加入服务器创建的 session
6	void setMaxInactiveInterval( )	设置两次请求间隔时间(单位是秒)

【例 4.6】 编写一个 JSP 页面 example4_6.jsp。如果用户是第一次访问该页面，会显示欢迎信息，并输出 session 对象允许的最长发呆时间、创建时间，以及 session 对象的 ID。在 example4_6.jsp 页面中，session 对象使用 setMaxInactiveInterval(int maxValue)方法设置最长的"发呆"状态时间为 20 秒。用户如果两次刷新间隔时间超过 20 秒，用户先前的 session 被取消，用户将获得一个新的 session 对象。页面运行效果如图 4.7(a)和图 4.7(b)所示。

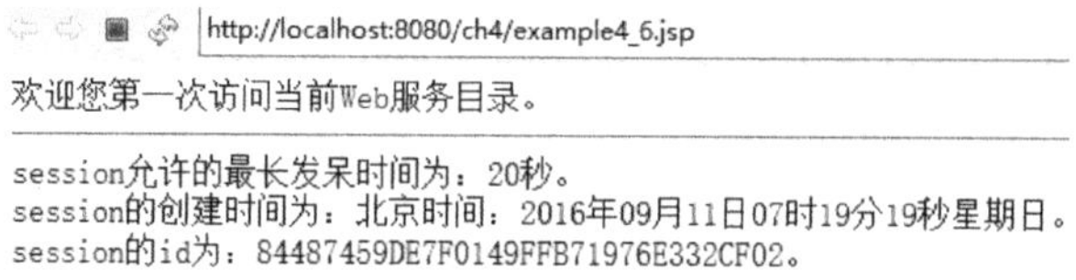

(a) 第一次或间隔20秒后访问该页面

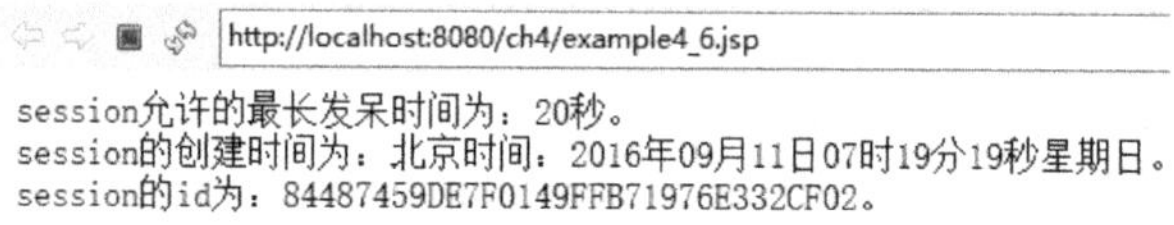

(b) 20秒之内访问该页面

图 4.7　session 生存期限

例 4.6 页面文件 example4_6.jsp 的代码如下：

```jsp
<%@ page language = "java" contentType = "text/html; charset = GBK" pageEncoding = "GBK" %>
<%@ page import = "java.util.*" %>
<%@ page import = "java.text.*" %>
<html>
    <head>
        <meta http-equiv = "Content-Type" content = "text/html; charset = ISO-8859-1">
        <title>example4_6.jsp</title>
    </head>
    <body>
    <%
//session 调用 setMaxInactiveInterval(int n)方法设置最长"发呆"时间为 20 秒
session.setMaxInactiveInterval(20);
//session 调用 isNew()方法判断 session 是不是新创建的
        boolean flg = session.isNew();
        if (flg) {
            out.println("欢迎您第一次访问当前 Web 服务目录.");
            out.println("<hr/>");
        }
        out.println("session 允许的最长发呆时间为：" +
        session.getMaxInactiveInterval() + "秒。");
        //获取 session 对象被创建的时间
        long num = session.getCreationTime();
        //将整数转换为 Date 对象
```

```
            Date time = new Date(num);
            //用给定的模式和默认语言环境的日期格式符号构造 SimpleDateFormat 对象
            SimpleDateFormat matter = new SimpleDateFormat(
                    "北京时间：yyyy 年 MM 月 dd 日 HH 时 mm 分 ss 秒 E。");
            //得到格式化后的字符串
            String strTime = matter.format(time);
            out.println("<br/>session 的创建时间为：" + strTime);
            out.println("<br/>session 的 id 为：" + session.getId() + "。");
        %>
    </body>
</html>
```

从例 4.6 中可以看出，如果用户长时间不关闭浏览器，session 对象也没有调用 invalidate()方法，那么用户的 session 也可能消失。例如，如果该例中的 JSP 页面在 20 秒之内不被访问，它先前创建的 session 对象就消失了，服务器又重新创建一个 session 对象。这是因为 session 对象达到了它的最大"发呆"时间。所谓"发呆"状态时间，是指用户对某个 Web 服务目录发出的两次请求之间的间隔时间。

用户对某个 Web 服务目录下的 JSP 页面发出请求并得到响应，如果用户不再对该 Web 服务目录发出请求，比如不再操作浏览器，那么用户对该 Web 服务目录进入"发呆"状态，直到用户再次请求该 Web 服务目录时，"发呆"状态结束。

Tomcat 服务器允许用户最长的"发呆"状态时间为 30 分钟。可以通过修改 Tomcat 安装目录中 conf 文件夹下的配置文件 web.xml，找到下面的片段，修改其中的默认值"30"，就可以重新设置各个 Web 服务目录下的 session 对象的最长"发呆"时间。这里的时间单位为分。

```
<session-config>
    <session-timeout>30</session-timeout>
</session-config>
```

也可以通过 session 对象调用 setMaxInactiveInterval(int time)方法来设置最长"发呆"状态时间，参数的时间单位为秒。

4.3.2 能力目标

理解 session 对象的生存期限，灵活使用 session 对象存储数据。

4.3.3 任务驱动

1. 任务的主要内容

编写 3 个 JSP 页面(task4_3_1.jsp、task4_3_2.jsp 和 task4_3_3.jsp)模拟登录系统的权限控制。在 task4_3_1.jsp 页面中输入用户名(test)和密码(test)，提交给 task4_3_2.jsp 页面。在 task4_3_2.jsp 页面中判断用户名和密码是否正确，正确则将用户名和密码存储在 session 对象中。没有成功登录，直接访问 task4_3_3.jsp 页面时，提示没有权限访问该页面。页面运行效果如图 4.8 所示。

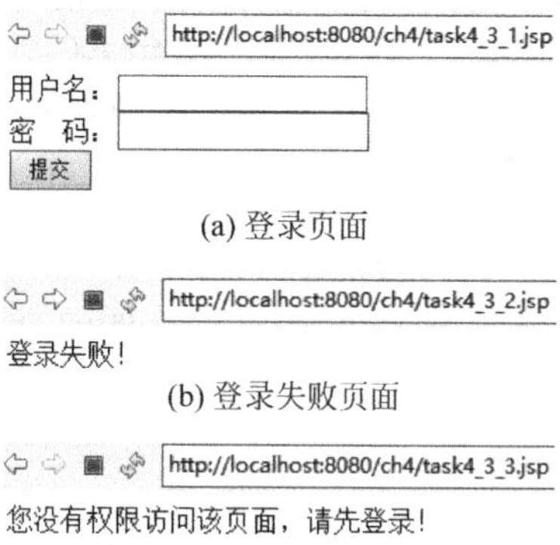

(a) 登录页面

(b) 登录失败页面

(c) 未成功登录直接访问该页面

图 4.8　权限模拟

2. 任务的代码模板

task4_3_1.jsp 的代码如下：

```jsp
<%@ page language="java" contentType="text/html; charset=GBK" pageEncoding="GBK"%>
<html>
    <head>
        <title>task4_3_1.jsp</title>
    </head>
    <body>
        <form action="task4_3_2.jsp" method="post">
            用户名：<input type="text" name="userName"/><br>
            密  码：<input type="password" name="pwd"/><br>
            <input type="submit" value="提交"/>
        </form>
    </body>
</html>
```

task4_3_2.jsp 的代码模板如下：

```jsp
<%@ page language="java" contentType="text/html; charset=GBK" pageEncoding="GBK"%>
<html>
    <head>
        <title>task4_3_2.jsp</title>
    </head>
    <body>
        <%
            String name = request.getParameter("userName");
            String pwd = request.getParameter("pwd");
            if("test".equals(name) && "test".equals(pwd)){
                【代码1】//将用户名 name 以 username 为 key 存储在 session 对象中
                out.print("登录成功!");
            }else{
                【代码2】//session 调用 invalidate()方法使 session 失效
```

```
                out.print("登录失败!");
            }
        %>
    </body>
</html>
```

task4_3_3.jsp 的代码模板如下：

```
<%@ page language = "java" contentType = "text/html; charset = GBK" pageEncoding = "GBK" %>
<html>
    <head>
        <title>task4_3_3.jsp</title>
    </head>
    <body>
        <%
            Object name =【代码 3】//从 session 对象取出用户名 name
            if(name == null){
                out.print("您没有权限访问该页面,请先登录!");
            }else{
                out.print("成功登录后的操作!");
            }
        %>
    </body>
</html>
```

3. 任务小结或知识扩展

客户端与服务器进行通信的协议是 HTTP 协议,该协议本身是基于请求/响应模式的、无状态的协议,服务器不会记录客户端的任何信息,这样客户端每次发送的请求都是独立的,这样的方式在工程实践中是不可用的。而会话(session)正是一种能将客户端信息保存在服务器端的技术,它可以记录客户端到服务器的一系列请求。在实际工程中,一般使用 session 跟踪用户的状态。例如,用户登录成功后,将用户信息保存到 session 中。

4. 代码模板的参考答案

【代码 1】: session.setAttribute("username", name);
【代码 2】: session.invalidate();
【代码 3】: session.getAttribute("username");

4.3.4 实践环节

用户到便民超市采购商品,购物前需要先登录会员卡号,购物时先将选购的商品放入购物车,最后到柜台清点商品。请借助于 session 对象模拟购物车,并存储客户的会员卡号和购买的商品名称。会员卡号输入后可以修改,购物车中的商品可以查看。编写程序模拟上述过程。loginID.jsp 实现会员卡号输入,shop.jsp 实现商品导购,food.jsp 实现商品购物,count.jsp 实现清点商品。本节实践环节的 4 个 JSP 页面都保存在目录 practice4 中,先运行 loginID.jsp 页面,运行效果如图 4.9 所示。

图 4.9 session 模拟购物车

4.4 全局应用程序对象 application

4.4.1 核心知识

不同用户的 session 对象互不相同,但有时候用户之间可能需要共享一个对象,Web 服务器启动后,就产生了这样一个唯一的内置对象 application。任何用户在访问同一 Web 服务目录各个页面时,共享一个 application 对象,直到服务器关闭,这个 application 对象才被取消。

application 同 session 对象一样也可以进行数据存储,处理数据方法如下。

(1) public void setAttribute(String key,Object obj):将参数 obj 指定的对象保存到 application 对象中,key 为所保存的对象指定一个关键字。若保存的两个对象关键字相同,则先保存的对象被清除。

(2) public Object getAttribute(String key):获取 application 中关键字是 key 的对象。

(3) public void removeAttribute(String key):从 application 中删除关键字 key 所对应的对象。

(4) public Enumeration getAttributeNames():产生一个枚举对象,该枚举对象可使用方法 nextElemets() 遍历 application 中的各个对象所对应的关键字。

4.4.2 能力目标

理解 application 对象的生存期限，灵活使用 application 对象存储数据。

4.4.3 任务驱动

1. 任务的主要内容

用 application 制作"成语接龙"，用户通过 task4_4_1.jsp 向 task4_4_2.jsp 提交四字成语，task4_4_2.jsp 页面获取成语内容后，用同步方法将该成语内容和以前的成语内容进行连接，然后将这些四字成语内容添加到 application 对象中。页面运行效果如图 4.10 所示。

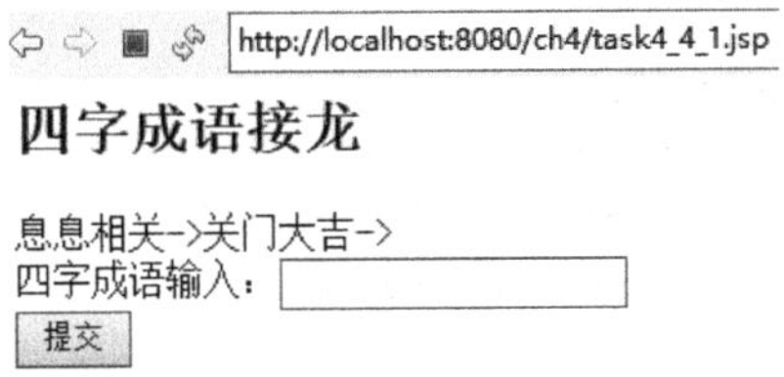

四字成语接龙

息息相关->关门大吉->
四字成语输入：
提交

(a) 成语提交页面

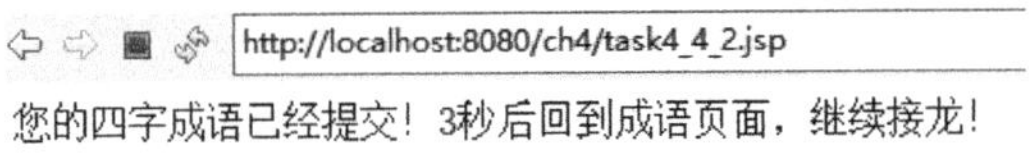

您的四字成语已经提交！3秒后回到成语页面，继续接龙！

(b) 接龙成功页面

图 4.10 成语接龙

2. 任务的代码模板

task4_4_1.jsp 的代码模板如下：

```jsp
<%@ page language="java" contentType="text/html; charset=GBK" pageEncoding="GBK"%>
<html>
    <head>
        <title>task4_4_1.jsp</title>
    </head>
    <body>
        <h2>四字成语接龙</h2>
        <%
            String s = 【代码1】//取出 application 中关键字是 message 的对象(成语内容)
            if(s! = null){
                out.print(s);
            }
            else{
                out.print("还没有词语,请您龙头开始!<br>");
            }
        %>
        <form action="task4_4_2.jsp" method="post">
            四字成语输入：<input type="text" name="mes"/><br>
            <input type="submit" value="提交"/>
        </form>
```

```
                </body>
    </html>
```

task4_4_2.jsp 的代码模板如下：

```
<%@ page language = "java" contentType = "text/html; charset = GBK" pageEncoding = "GBK" %>
<%@ page import = "java.util.*" %>
<html>
    <head>
        <title>task4_4_2.jsp</title>
    </head>
    <body>
        <%!
            String message = "";
            ServletContext application;
            synchronized void sendMessage(String s){
                application = getServletContext();
                message = message + s + " ->";
                【代码 2】//把成语内容 message 以 message 为关键字存储到 application 对象中
            }
        %>

        <%
            request.setCharacterEncoding("GBK");
            String content = request.getParameter("mes");
            sendMessage(content);
            out.print("您的四字成语已经提交!3 秒后回到成语页面,继续接龙!");
            response.setHeader("refresh", "3;url = task4_4_1.jsp");
        %>
    </body>
</html>
```

3. 任务小结或知识扩展

任务中成语接龙方法 sendMessage 为什么定义为同步方法呢？这是因为 application 对象对所有的用户都是相同的,任何用户对该对象中存储的数据的操作都会影响其他用户。

如果客户端浏览不同的 Web 服务目录,将产生不同的 application 对象。同一个 Web 服务目录中的所有 JSP 页面都共享同一个 application 对象,即使浏览这些 JSP 页面的客户不相同也是如此。因此,保存在 application 对象中的数据不仅可以跨页面分享,还可以由所有用户共享。

有些 Web 服务器不能直接使用 application 对象,必须使用父类 ServletContext 声明这个对象,然后使用 getServletContext()方法为 application 对象进行实例化。例如,该任务中 task4_4_2.jsp 页面中的代码。

4. 代码模板的参考答案

【代码 1】: (String)application.getAttribute("message");
【代码 2】: application.setAttribute("message", message);

4.4.4 实践环节

使用 application 对象实现网站访客计数器的功能。计数器运行效果如图 4.11 所示。

图 4.11 计数器页面

4.5 小　　结

本章重点介绍了 request、session 和 application 对象，三者区别如下。

request 对象内数据的存活范围就是在 request 对象的存活范围内，当客户端向服务器端发送一个请求，服务器向客户端返回一个响应后，该请求对象就被销毁了；之后再向服务器端发送新的请求时，服务器会创建新的 request 对象，该 request 对象与之前的 request 对象没有任何关系，因此也无法获得在之前的 request 对象中所存放的任何数据。

session 是服务器端的行为，用于跟踪客户的状态，当用户去访问某个站点时，服务器端就会为客户产生一个 sessionID，以 cookie 的方式返回给客户端，当客户去访问该站点的其他服务时，就会带着当前 sessionID 一起发出请求，以识别是哪个用户，一个用户就好比一个 session 对象，互不干扰。

session 对象内数据的存活范围也就是 session 对象的存活范围（只要浏览器不关闭，session 对象就会一直存在），因此在同一个浏览器窗口中，无论向服务器端发送多少个请求，session 对象只有一个。session 对象经常用在登录和购物车场景。

application 对象是存活范围最大的对象，只要服务器没有关闭，application 对象中的数据就会一直存在。在整个服务器运行过程中，application 对象只有一个。所有用户共享这个 application 对象。该对象经常用在统计网站访问次数场景。

request、session 以及 application 这 3 个对象的范围是逐个增加的：request 只在一个请求的范围内；session 是在浏览器窗口的范围内；application 则是在整个服务器的运行过程中。

习　题　4

1. 下面(　　)操作不能关闭 session 对象。

　　A. 用户刷新当前页面调用

　　B. 用户关闭浏览器

　　C. session 达到设置的最长"发呆"时间

　　D. session 对象的 invalidate()方法

2. 有如下程序片段：

```
<form>
```

```
< input type = "text" name = "id">
< input type = "submit" value = "提交">
</form >
```

下面(　　)语句可以获取用户输入的信息。

  A. request.getParameter("id");

  B. request.getAttribute("submit");

  C. session.getParameter(key，"id");

  D. session.getAttribute(key，"id");

3. 下面(　　)内置对象是对客户的请求做出响应,向客户端发送数据的。

  A. request

  B. session

  C. response

  D. application

4. 可以使用(　　)方法实现客户的重定向。

  A. response.setStatus();

  B. response.setHeader();

  C. response.setContentType();

  D. response.sendRedirect();

5. 什么对象是内置对象？常见的内置对象有哪些？

6. 请简述内置对象 request、session 和 application 之间的区别。

7. 一个用户在不同 Web 服务目录中的 session 对象相同吗？一个用户在同一 Web 服务目录的不同子目录中的 session 对象相同吗？

8. session 对象的生存期限依赖于哪些因素？

9. 简述 forward 动作标记与 response.sendRedirect()两种跳转的区别。

第 5 章

JSP 与 JavaBean

主要内容

(1) JavaBean 的含义。
(2) JSP 中使用 JavaBean。

一个 JSP 页面通过使用 HTML 标记为用户显示数据(静态部分),页面中变量的声明、程序片以及表达式为动态部分,对数据进行处理。如果 Java 程序片和 HTML 标记大量掺杂在一起使用,就不利于 JSP 页面的扩展和维护。JSP 和 JavaBean 技术的结合不仅可以实现数据的表示和处理分离,而且可以提高 JSP 程序代码重用的程度,是 JSP 编程中常用的技术。

本章涉及的 Java 源文件保存在工程 ch5 的 src 中,涉及的 JSP 页面保存在工程 ch5 的 WebContent 中。

5.1 编写 JavaBean

5.1.1 核心知识

JavaBean 是一个可重复使用的软件组件,是遵循一定标准、用 Java 语言编写的一个类,该类的一个实例称为一个 JavaBean,简称 bean。JavaBean 具有可重用,升级方便,不依赖于平台等特点。JavaBean 又可分为业务 bean 和数据 bean。业务 bean 用于封装业务逻辑、数据库操作等;数据 bean 用于封装数据。

编写一个 JavaBean 就是编写一个 Java 类(该类必须带有包名),这个类创建的一个对象称为一个 bean,为了让 JSP 引擎(比如 Tomcat)知道这个 bean 的属性和方法,JavaBean 类必须遵守以下规则。

(1) 如果类的成员变量的名字是 name,那么为了获取或更改成员变量的值,类中必须提供两个方法:

getName(),用来获取属性 name。
setName(),用来修改属性 name。
即方法的名字用 get 或 set 为前缀,后缀是首字母大写的成员变量的名字。

(2) 对于 boolean 类型的成员变量,允许使用 is 代替上面的 get 和 set。

（3）类中方法的访问权限必须是 public。

（4）构造方法必须无参数。

5.1.2 能力目标

能够灵活使用 JavaBean 的编写规则编写创建 bean 的 Java 源文件。

5.1.3 任务驱动

1. 任务的主要内容

创建 bean 的源文件 Rectangle.java（在包 small.dog 中），该 bean 的作用是计算矩形的面积和周长。

2. 任务的代码模板

Rectangle.java 的代码模板如下：

```java
package com.bean;
public class Rectangle {
    private double length;
    private double width;
    【代码 1】{                    //定义类 Rectangle 的构造方法
            length = 20;
            width = 10;
    }
    【代码 2】{                    //定义获取矩形长度的方法
        return length;
    }
    【代码 3】{                    //定义修改矩形长度的方法
        this.length = length;
    }
    public double getWidth() {
        return width;
    }
    public void setWidth(double width) {
        this.width = width;
    }
    public double computerArea(){
        return length * width;
    }
    public double computerLength(){
        return (length + width) * 2;
    }
}
```

3. 任务小结或知识扩展

JavaBean 可以在任何 Java 程序编写环境下完成编写，再通过编译成为一个字节码文件，为了让 JSP 引擎（比如 Tomcat）找到这个字节码，必须把字节码文件存放在特定的位置。本书使用 Eclipse 集成环境开发 JSP 程序，Java 类的字节码文件由 Eclipse 自动保存到 Web 工程的 build\classes 中。例如，本任务中的 Rectangle.class 文件保存在 ch5\build\classes\

com\bean 目录中。

JavaBean 是基于 Java 语言的,因此 JavaBean 具有以下特点。

（1）与平台无关。

（2）代码可重复利用。

（3）易扩展、易维护、易使用。

4. 代码模板的参考答案

【代码 1】: public Rectangle()
【代码 2】: public double getLength()
【代码 3】: public void setLength(double length)

5.1.4 实践环节

创建 bean 的源文件 Circle.java(在包 com.bean 中),该 bean 的作用是计算圆的面积和周长。

5.2 JSP 中使用 JavaBean

在 JSP 页面中使用 bean 时,首先使用 page 指令的 import 属性导入创建 bean 的类,例如:

<%@ page import = "com.bean.*" %>

5.2.1 核心知识

1. 动作标记 useBean

useBean 动作标记用来查找或者实例化一个 JavaBean。useBean 标记的格式如下:

<jsp:useBean id = "bean 的名字" class = "创建 bean 的类" scope = "bean 的有效范围" />

或

<jsp:useBean id = "bean 的名字" type = "创建 bean 的类" scope = "bean 的有效范围" />

例如:

<jsp:useBean id = "rectangle" class = "com.bean.Rectangle" scope = "page"/>

useBean 标记中各属性含义如表 5.1 所示。

表 5.1 useBean 标记属性含义

属性名	描 述
id	指定该 JavaBean 实例的变量名,通过 id 可以访问这个实例
class	指定 JavaBean 的类名。如果需要创建一个新的实例,Web 容器会使用 class 指定的类,并调用无参数的构造方法来完成实例化

续表

属性名	描 述
scope	指定 JavaBean 的作用范围,包括 page、request、session 和 application。默认值为 page,表明此 JavaBean 只能应用于当前页;值为 request 表明此 JavaBean 只能应用于当前的请求;值为 session 表明此 JavaBean 能应用于当前会话;值为 application 则表明此 JavaBean 能应用于整个应用程序内
type	指定 JavaBean 对象的类型,通常在查找已存在的 JavaBean 时使用,这时使用 type 将不会产生新的对象 如果是查找已存在的 JavaBean 对象,type 属性的值可以是此对象的准确类名、其父类或者其实现的接口;如果是新建实例,则只能是准确类名或者父类 另外,如果能够确定此 JavaBean 的对象肯定存在,则指定 type 属性后可以省略 class 属性

当含有 useBean 动作标记的 JSP 页面被 Web 容器加载执行时,Web 容器首先根据 id 的名字,在 pageContent 内置对象中查看是否含有名字为 id 和作用域为 scope 的对象;如果该对象存在,Web 容器就将这个对象的副本(bean)分配给 JSP 页面使用;如果没有找到,就根据 class 指定的类创建一个名字是 id 的 bean,并添加到 pageContent 对象中,同时将这个 bean 分配给 JSP 页面使用。useBean 动作标记执行流程如图 5.1 所示。

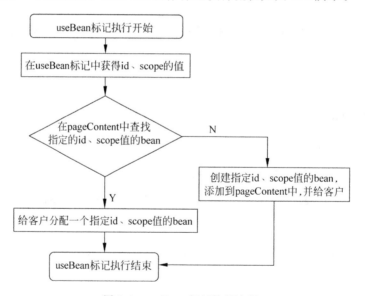

图 5.1 useBean 标记执行流程

【例 5.1】 编写一个 JSP 页面 example5_1.jsp,在 JSP 页面中使用 useBean 标记获得一个 bean,负责创建 bean 的类是 5.1.3 节任务中的 Rectangle 类,bean 的名字是 rectangle,rectangle 的 scope 取值为 page。JSP 页面的运行效果如图 5.2 所示。

例 5.1 页面文件 example5_1.jsp 的代码如下:

```
<%@ page language = "java" contentType = "text/html; charset = GBK" pageEncoding = "GBK" %>
<%@ page import = "com.bean.Rectangle" %>
<html>
    <head>
```

```
                <title>example5_1.jsp</title>
        </head>
        <body>
            <jsp:useBean id = "rectangle" class = "com.bean.Rectangle" scope = "page"/>
            <p>矩形的长是：<% = rectangle.getLength() %>
            <p>矩形的宽是：<% = rectangle.getWidth() %>
            <p>矩形的面积是：<% = rectangle.computerArea() %>
            <p>矩形的周长是：<% = rectangle.computerLength() %>
        </body>
</html>
```

2. 动作标记 getProperty

JavaBean 的实质是遵守一定规范的类所创建的对象，可以通过两种方式获取 bean 的属性：①在 Java 程序片或表达式中，使用 bean 对象调用 getXxx()方法获取 bean 的属性值，比如，例 5.1 中的语句：<% = rectangle.getLength()% >；②先通过<jsp:useBean>标记获得一个 bean，再通过<jsp:getProperty>标记获取 bean 的属性值。

使用 getProperty 动作标记可以获得 bean 的属性值。使用该动作标记之前，必须事先使用 useBean 动作标记获得一个相应的 bean。getProperty 动作标记语法格式如下：

`<jsp:getProperty name = "bean 的名字" property = "bean 的属性" />`

或

`<jsp:getProperty name = "bean 的名字" property = "bean 的属性"/></jsp:getProperty>`

其中，name 取值是 bean 的名字，和 useBean 标记中的 id 对应；property 取值是 bean 的一个属性名字，和创建该 bean 的类的成员变量名对应。这条指令相当于在 Java 表达式中使用 bean 的名字调用 getXxx()方法。

【例 5.2】 创建 bean 的源文件 NewRectangle.java，该 bean 的作用是计算矩形的面积和周长。编写一个 JSP 页面 useGetProperty.jsp，在该 JSP 页面中使用 useBean 标记创建一个名字是 pig 的 bean，并使用 getProperty 动作标记获得 pig 的每个属性值。负责创建 pig 的类是 NewRectangle 类。JSP 页面运行效果如图 5.3 所示。

NewRectangle.java 的代码如下：

```
package com.bean;
public class NewRectangle {
    double length;
    double width;
```

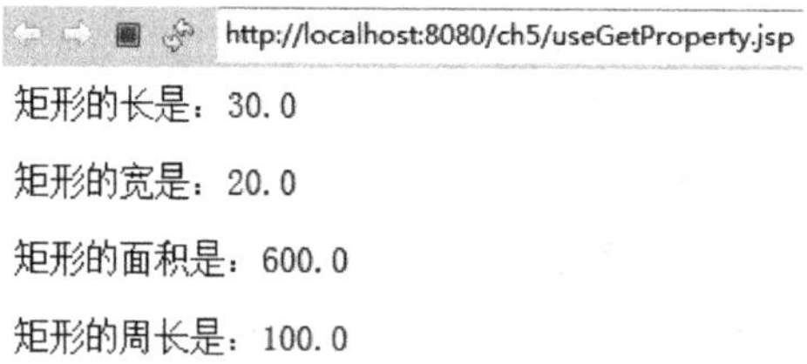

矩形的长是：30.0

矩形的宽是：20.0

矩形的面积是：600.0

矩形的周长是：100.0

图 5.3　使用 getProperty 标记获得 bean 的属性值

```java
    double rectangleArea;
    double rectangleLength;
    public NewRectangle() {
        length = 20;
        width = 10;
    }
    public double getLength() {
        return length;
    }
    public void setLength(double length) {
        this.length = length;
    }
    public double getWidth() {
        return width;
    }
    public void setWidth(double width) {
        this.width = width;
    }
    public double getRectangleArea() {
        return length * width;
    }
    public double getRectangleLength() {
        return 2 * (width + length);
    }
}
```

例 5.2 页面文件 useGetProperty.jsp 的代码如下：

```jsp
<%@ page language = "java" contentType = "text/html; charset = GBK" pageEncoding = "GBK" %>
<%@ page import = "com.bean.NewRectangle" %>
<html>
    <head>
        <title>useGetProperty.jsp</title>
    </head>
    <body>
        <jsp:useBean id = "pig" class = "com.bean.NewRectangle" scope = "page"/>
        <% pig.setLength(30); %>
        <% pig.setWidth(20); %>
        <p>矩形的长是：<jsp:getProperty property = "length" name = "pig"/>
        <p>矩形的宽是：<jsp:getProperty property = "width" name = "pig"/>
        <p>矩形的面积是：<jsp:getProperty property = "rectangleArea" name = "pig"/>
        <p>矩形的周长是：<jsp:getProperty property = "rectangleLength" name = "pig"/>
```

            </body>
        </html>

3. 动作标记 setProperty

除了在 Java 程序片中使用 bean 对象调用 setXxx()方法修改 bean 的属性值外,可以使用 setProperty 动作标记修改 bean 的属性值。使用该动作标记之前,必须事先使用 useBean 动作标记获得一个相应的 bean。使用 setProperty 动作标记进行 bean 属性值的设置有以下 3 种方式。

1)用表达式或字符串设置 bean 的属性

① 用表达式设置 bean 的属性。

<jsp:setProperty name="bean 的名字" property="bean 的属性" value="<%=expression%>" />

② 用字符串设置 bean 的属性。

<jsp:setProperty name="bean 的名字" property="bean 的属性" value="字符串" />

用表达式修改 bean 属性值时,表达式值的类型必须与 bean 的属性类型一致。用字符串修改 bean 属性值时,字符串会自动被转化为 bean 的属性类型,不能转化成功的可能会抛出 NumberFormatException 异常。

2)通过 HTTP 表单参数值设置 bean 的属性

<jsp:setProperty name="bean 的名字" property="*" />

通过 HTTP 表单参数值设置 bean 的属性时,表单参数的名字必须与 bean 属性的名字相同,服务器会根据名字自动匹配,类型自动转换。

3)任意指定请求参数设置 bean 的属性

<jsp:setProperty name="bean 的名字" property="属性名" param="参数名"/>

可以根据自己的需要,任意选择传递的参数,请求参数名无须与 bean 属性名相同。

【例 5.3】 用表达式或字符串修改 bean 的属性。具体要求如下。

(1)创建 bean 的源文件 Car.java,该 bean 的作用是描述小汽车的一些属性。

(2)编写一个 JSP 页面 car.jsp,在该 JSP 页面中使用 useBean 标记创建一个名字为 smallCar 的 bean,其有效范围是 page,并使用动作标记修改、获取该 bean 的属性值。负责创建 smallCar 的类是 Car。JSP 页面运行效果如图 5.4 所示。

图 5.4 使用字符串或表达式修改 bean 的属性

Car.java 的代码如下:

```
package com.bean;
public class Car {
    String tradeMark;
    String number;
```

```
    public String getTradeMark() {
        return tradeMark;
    }
    public void setTradeMark(String tradeMark) {
        this.tradeMark = tradeMark;
    }
    public String getNumber() {
        return number;
    }
    public void setNumber(String number) {
        this.number = number;
    }
}
```

例 5.3 页面文件 car.jsp 的代码如下：

```
<%@ page language="java" contentType="text/html; charset=GBK" pageEncoding="GBK"%>
<%@ page import="com.bean.Car"%>
<html>
    <head>
        <title>car.jsp</title>
    </head>
    <body>
        <jsp:useBean id="smallCar" class="com.bean.Car" scope="page"/>
        <%
            String carNo = "京A8888";
        %>
        <%-- 使用setProperty标记设置smallCar的属性tradeMark值为"宝马X6" --%>
        <jsp:setProperty property="tradeMark" name="smallCar" value="宝马X6"/>
        <%-- 使用setProperty标记设置smallCar的属性number值为carNo --%>
        <jsp:setProperty property="number" name="smallCar" value="<%=carNo%>"/>
        汽车的品牌是：<jsp:getProperty property="tradeMark" name="smallCar"/>
        <br>汽车的牌号是：<jsp:getProperty property="number" name="smallCar"/>
    </body>
</html>
```

【例 5.4】 通过 HTTP 表单参数值设置 bean 的属性。具体要求如下。

(1) 编写 JSP 页面 inputCar.jsp 和 showCar.jsp。

(2) 在 inputCar.jsp 页面中输入信息后提交给 showCar.jsp 页面显示信息。

(3) JSP 页面中用到的 bean 是例 5.3 中 Car 类创建的。JSP 页面运行效果如图 5.5 所示。

(a) 信息输入页面

(b) 信息显示页面

图 5.5　运行效果

例 5.4 页面文件 inputCar.jsp 的代码如下：

```
<%@ page language="java" contentType="text/html; charset=GBK" pageEncoding="GBK"%>
<html>
    <head>
        <title>inputCar.jsp</title>
    </head>
    <body>
        <form action="showCar.jsp" method="post">
            请输入汽车品牌：
            <input type="text" name="tradeMark"/>
            <br>
            请输入汽车牌号：
            <input type="text" name="number"/>
            <br>
            <input type="submit" value="提交"/>
        </form>
    </body>
</html>
```

例 5.4 页面文件 showCar.jsp 的代码如下：

```
<%@ page language="java" contentType="text/html; charset=GBK" pageEncoding="GBK"%>
<%@ page import="com.bean.Car"%>
<%
request.setCharacterEncoding("GBK");
%>
<html>
    <head>
        <title>showCar.jsp</title>
    </head>
    <body>
        <jsp:useBean id="smallCar" class="com.bean.Car" scope="page"/>
        <%-- 通过HTTP表单的参数的值设置bean的属性(表单参数与属性自动匹配) --%>
        <jsp:setProperty property="*" name="smallCar"/>
        汽车的品牌是：<jsp:getProperty property="tradeMark" name="smallCar"/>
        <br>汽车的牌号是：<jsp:getProperty property="number" name="smallCar"/>
    </body>
</html>
```

5.2.2 能力目标

能够在 JSP 页面中灵活使用动作标记 useBean、getProperty 和 setProperty。

5.2.3 任务驱动

1. 任务的主要内容

编写两个 JSP 页面：login.jsp 与 invalidate.jsp。login.jsp 页面提供一个表单，用户通过表单将用户名和密码（正确的用户名和密码分别是 zhangsan 和 123456）提交给 invalidate.jsp 页面。用户提交表单后，JSP 页面将登录验证的任务提交给一个 bean 去完成。页面运行效果如图 5.6 所示。

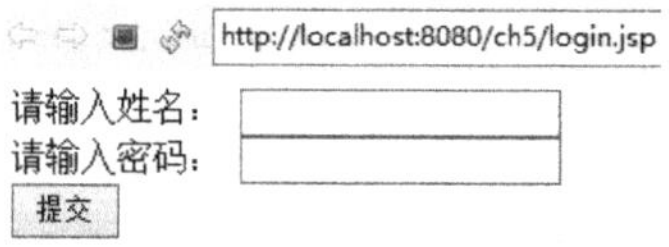

(a) 信息输入页面

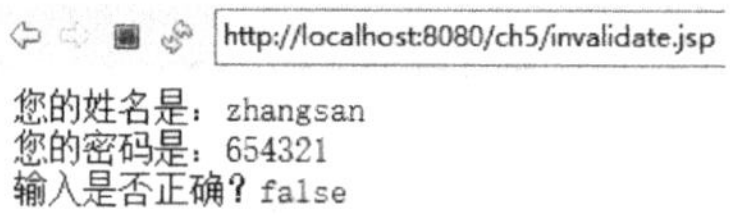

(b) 登录验证页面

图 5.6 页面运行效果

2. 任务的代码模板

LoginBean.java 的代码如下:

```java
package com.bean;
public class LoginBean {
    String uname;
    String upass;
    boolean login;
    public String getUname() {
        return uname;
    }
    public void setUname(String uname) {
        this.uname = uname;
    }
    public String getUpass() {
        return upass;
    }
    public void setUpass(String upass) {
        this.upass = upass;
    }
    public boolean isLogin() {
        if(uname.equals("zhangsan")&&upass.equals("123456")){
            login = true;
        }else{
            Login = false;
        }
        return login;
    }
    public void setLogin(boolean login) {
        this.login = login;
    }
}
```

login.jsp 的代码如下:

```
<%@ page language = "java" contentType = "text/html; charset = GBK" pageEncoding = "GBK" %>
<html>
```

```
        <head>
            <title>login.jsp</title>
        </head>
        <body>
            <form action = "invalidate.jsp" method = "post">
                请输入姓名：
                <Input type = text name = "uname"/><BR>
                请输入密码：
                <Input type = text name = "upass"/><BR>
                <INPUT TYPE = "submit" value = "提交"/>
            </form>
        </body>
</html>
```

invalidate.jsp 的代码模板如下：

```
<%@ page language = "java" contentType = "text/html; Charset = GBK" pageEncoding = "GBK" %>
<%@ page import = "com.bean.LoginBean" %>
<%
request.setCharacterEncoding("GBK");
%>
<html>
    <head>
        <title>invalidate.jsp</title>
    </head>
    <body>
        【代码 1】  <%-- 使用 useBean 标记创建 bean 对象 lb,有效范围为 page --%>
        【代码 2】  <%-- 通过 HTTP 表单参数值设置 bean 的属性(表单参数与属性自动匹配) --%>
        您的姓名是：【代码 3】  <%-- 使用 getProperty 标记获得 bean 的属性值 --%>
        <br>您的密码是：【代码 4】  <%-- 使用 getProperty 标记获得 bean 的属性值 --%>
        <br>输入是否正确?【代码 5】  <%-- 使用 getProperty 标记获得 bean 的属性值 --%>
    </body>
</html>
```

3. 任务小结或知识扩展

如果 bean 的属性为 boolean 类型，可以使用 isXxx 代替 getXxx 方法，如本节任务中的 LoginBean 的 login。

4. 代码模板的参考答案

【代码 1】：<jsp:useBean id = "lb" class = "com.bean.LoginBean" scope = "page"/>
【代码 2】：<jsp:setProperty property = "*" name = "lb"/>
【代码 3】：<jsp:getProperty property = "uname" name = "lb"/>
【代码 4】：<jsp:getProperty property = "upass" name = "lb"/>
【代码 5】：<jsp:getProperty property = "login" name = "lb"/>

5.2.4 实践环节

编写两个 JSP 页面：inputTriangle.jsp 与 showTriangle.jsp。inputTriangle.jsp 提供一个表单，用户可以通过表单输入三角形的 3 条边提交给 showTriangle.jsp。用户提交表

单后,JSP 页面将计算三角形面积和周长的任务交给一个 bean 去完成,创建 bean 的源文件是 Triangle.java。页面运行效果如图 5.7 所示。

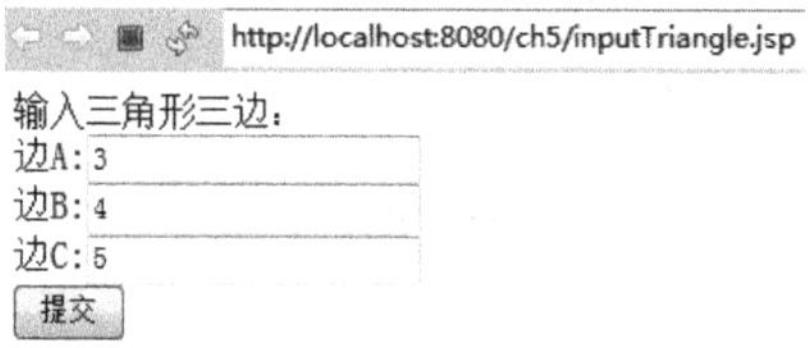

(a) 三角形边长输入页面

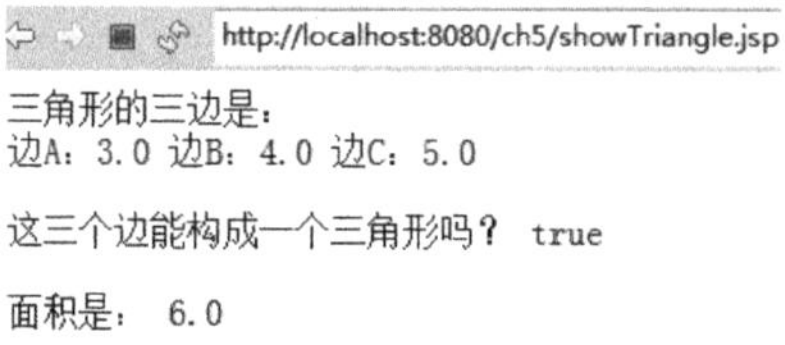

(b) 信息显示页面

图 5.7 页面运行效果

5.3 小　　结

本章重点介绍了在 JSP 中如何使用动作标记 useBean、getProperty 和 setProperty。

习　题　5

1. 下面(　　)是正确使用 JavaBean 的方式。
   A. <jsp:useBean id="address" class="tom.AddressBean" scope="page"/>
   B. <jsp:useBean name="address" class="tom.AddressBean" scope="page"/>
   C. <jsp:useBean bean="address" class="tom.AddressBean" scope="page"/>
   D. <jsp:useBean beanName="address" class="AddressBean" scope="page" />
2. JavaBean 中方法的访问属性必须是(　　)。
   A. private　　　　B. public　　　　C. protected　　　　D. friendly
3. 在 JSP 中调用 JavaBean 时不会用到的标记是(　　)。
   A. <javabean>　　　　　　　　　　B. <jsp:useBean>
   C. <jsp:setProperty>　　　　　　　D. <jsp:getProperty>
4. JavaBean 的作用域可以是(　　)、page、session 和 application。
   A. request　　　　B. response　　　　C. out　　　　D. 以上都不对
5. 在 test.jsp 文件中有如下一行代码:
   <jsp:useBean class="tom.jiafei.Test" id="user" scope="_____" />要使 user 对象一直存在于会话中,直至终止或被删除为止,下划线中应填入(　　)。
   A. page　　　　B. request　　　　C. session　　　　D. application
6. 在 JSP 中,使用<jsp:useBean>动作可以将 javaBean 引入 JSP 页面,对 JavaBean

的访问范围不能是（　　）。

　　A．page　　　　B．request　　　　C．response　　　　D．application

7．下面语句与<jsp:getProperty name="aBean" property="jsp"/>等价的是（　　）。

　　A．<%=jsp()%>

　　B．<%out.print(aBean.getJsp());%>

　　C．<%=aBean.setJsp()%>

　　D．<%aBean.setJsp();%>

8．以下是有关 jsp:setProperty 和 jsp:getProperty 标记的描述，正确的是（　　）。

　　A．<jsp:setProperty>和<jsp:getProperty>标记都必须在<jsp:useBean>的开始标记和结束标记之间

　　B．这两个标记的 name 属性值必须与<jsp:useBean>标记的 id 属性值相对应

　　C．这两个标记的 name 属性值可以与<jsp:userBean>标记的 id 属性值不同

　　D．以上均不对

9．在 JSP 中使用<jsp:getProperty>标记时，不会出现的属性是（　　）。

　　A．name　　　　　　　　　　　　B．property

　　C．value　　　　　　　　　　　　D．以上皆不会出现

JSP 访问数据库

主要内容

(1) 连接数据库的常用方式。
(2) 数据库操作。
(3) 游动查询。
(4) 访问 Excel 电子表格。
(5) 连接池。
(6) PreparedStatement 语句。

数据库在 Web 应用中扮演着越来越重要的作用。如果没有数据库，很多重要的应用，像电子商务、搜索引擎等都不可能实现。本章主要介绍在 JSP 中如何访问关系数据库，如 Oracle、SQL Server、MySQL 和 Microsoft Access 等。

本章涉及的 Java 源文件保存在工程 ch6 的 src 中，涉及的 JSP 页面保存在工程 ch6 的 WebContent 中。

6.1 使用 JDBC-ODBC 桥接器连接数据库

JSP 页面中访问数据库，首先要与数据库进行连接，通过连接向数据库发送指令，并获得返回的结果。JDBC 连接数据库有两种常用方式：建立 JDBC-ODBC 桥接器和加载纯 Java 驱动程序。

6.1.1 核心知识

JDBC(Java Database Connectivity)是用于运行 SQL 的解决方案，是 Java 运行平台核心类库中的一部分，它由一组标准接口与类组成。使用 JDBC 完成 3 件事：与指定的数据库建立连接；向已连接的数据库发送 SQL 命令；处理 SQL 命令返回的结果。

ODBC 是由 Microsoft 主导的数据库连接标准，提供了通用的数据库访问平台。但是，使用 ODBC 连接数据库的应用程序移植性较差，因为应用程序所在的计算机必须提供 ODBC。

使用 JDBC-ODBC 桥接器连接数据库的机制是，将连接数据库的相关信息提供给 JDBC-ODBC 驱动程序，然后转换成 JDBC 接口，供应用程序使用，而和数据库的连接由

ODBC 完成。使用 JDBC-ODBC 桥接器连接数据库的示意图如图 6.1 所示。

使用 JDBC-ODBC 桥接器连接数据库有以下 3 个步骤。

(1) 建立 JDBC-ODBC 桥接器。

(2) 创建 ODBC 数据源。

(3) 和 ODBC 数据源指定的数据库建立连接。

6.1.2　能力目标

掌握 JDBC-ODBC 桥接器连接数据库的方法。

6.1.3　任务驱动

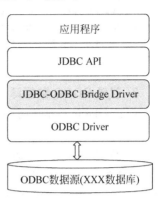

图 6.1　JDBC-ODBC 桥方式

任务的主要内容如下。

(1) 创建待连接的 Microsoft Access 数据库。

(2) 建立 JDBC-ODBC 桥接器。

(3) 创建 ODBC 数据源。

(4) 和 ODBC 数据源指定的数据库建立连接。

(5) 在 JSP 页面中使用 JDBC-ODBC 桥接器连接数据库。

1. 创建待连接的 Microsoft Access 数据库

使用 Microsoft Access 2007 设计一个数据库 goods.accdb，该库中有一张表，表的名字为 goodsInfo，表的说明如图 6.2 所示。

字段名称	数据类型	说明
goodsId	自动编号	商品编号
goodsName	文本	商品名称
goodsPrice	货币	商品价格
goodsType	文本	商品类型

图 6.2　goodsInfo 表的说明

表中的数据如图 6.3 所示。

goodsId	goodsName	goodsPric	goodsType	添加新字段
1	苹果	¥6.99	水果	
2	冰箱	¥8,999.00	电器	
3	热水壶	¥1,899.00	电器	
4	牛角面包	¥8.00	食品	
5	XXX上衣	¥888.00	服装	
6	牙膏	¥9.99	日用品	

图 6.3　goodsInfo 表中的数据

2. 建立 JDBC-ODBC 桥接器

JDBC 通过 java.lang.Class 类的静态方法 forName 加载 sun.jdbc.odbc.JdbcOdbcDriver 类建立 JDBC-ODBC 桥接器。建立桥接器时可能发生 ClassNotFoundException 异常，必须捕获该异常，建立桥接器的具体代码如下：

```
try{
    Class.forName("sun.jdbc.odbc.JdbcOdbcDriver");
}catch(ClassNotFoundException e){
    e.printStackTrace();
}
```

3. 创建 ODBC 数据源

创建 ODBC 数据源时,必须保证计算机有 ODBC 系统,Windows 操作系统一般都自带 ODBC 系统。

1) 打开 ODBC 数据源管理器

在 Windows 7 系统下,首先打开"控制面板"窗口中的"系统和安全"窗口,然后打开"管理工具"窗口,最后找到"数据源(ODBC)"图标双击打开,出现如图 6.4 所示的对话框。

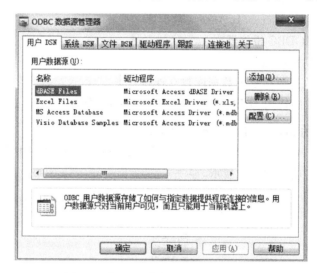

图 6.4 "ODBC 数据源管理器"对话框

2) 为数据源选择驱动程序

在图 6.4 所示的界面上选择"用户 DSN"选项卡,单击"添加"按钮,出现为新增的数据源选择驱动程序界面,如图 6.5 所示。因为要连接 Microsoft Access 2007 数据库,选择 Microsoft Access Driver(*.mdb,*.accdb)选项,单击"完成"按钮。如果在图 6.5 所示的界面中没有 Microsoft Access 的驱动程序,那么打开目录 C:\Windows\SysWOW64,双击该目录下的 odbcad32.exe 文件,打开"ODBC 数据源管理器"对话框,在此对话框中就有 Microsoft Access 的驱动了。

3) 为数据源起名并找到对应的数据库

在图 6.5 所示的界面中单击"完成"按钮,出现设置数据源具体信息的对话框,如图 6.6 所示。在"数据源名"文本框中为数据源起个名字,如 myGod。在图 6.6 所示的界面中单击"选择"按钮,为 myGod 数据源选择数据库。

4) 设置登录名和密码

在图 6.6 所示的界面中单击"高级"按钮,出现设置登录名与密码界面,如图 6.7 所示。这里的用户名和密码都是 firstDB。在图 6.7 所示的界面中单击"确定"按钮后,再单击

图 6.6 所示的界面中的"确定"按钮就创建了一个新的数据源：myGod。

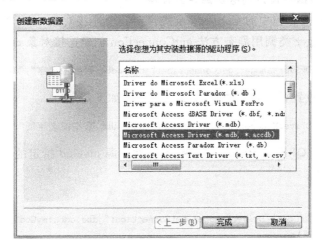

图 6.5 "创建新数据源"对话框

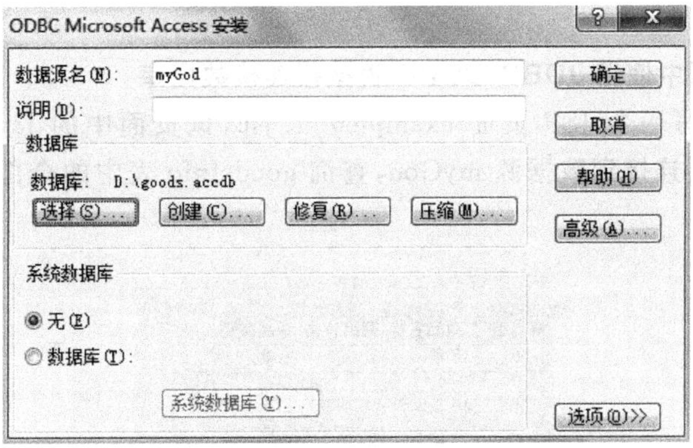

图 6.6 "ODBC Microsoft Access 安装"对话框

图 6.7 "设置高级选项"对话框

4. 和 ODBC 数据源指定的数据库建立连接

首先，使用 java. sql 包中的 Connection 类声明一个连接对象 con，然后再使用 java. sql 包中的 DriverManager 类调用静态方法 getConnection 创建连接对象 con：

```
Connection con = DriverManager.getConnection("jdbc:odbc:数据源名字","登录名","密码");
```

如果没有给数据源设置登录名和密码，那么连接形式如下：

```
Connection con = DriverManager.getConnection("jdbc:odbc:数据源名字","","");
```

建立连接时应捕获 SQLException 异常，例如，和数据源 myGod 指定的数据库 goods. accdb 建立连接，代码如下：

```
try{
        Connection con = DriverManager.getConnection("jdbc:odbc:myGod","firstDB",
        "firstDB");
}catch(SQLException e){
        e.printStackTrace();
}
```

5. 在 JSP 页面中使用 JDBC-ODBC 桥接器连接数据库

【例 6.1】 编写一个 JSP 页面 example6_1. jsp，该页面中的 Java 程序片代码使用 JDBC-ODBC 桥接器连接到数据源 myGod，查询 goodsInfo 表中的全部记录。页面运行效果如图 6.8 所示。

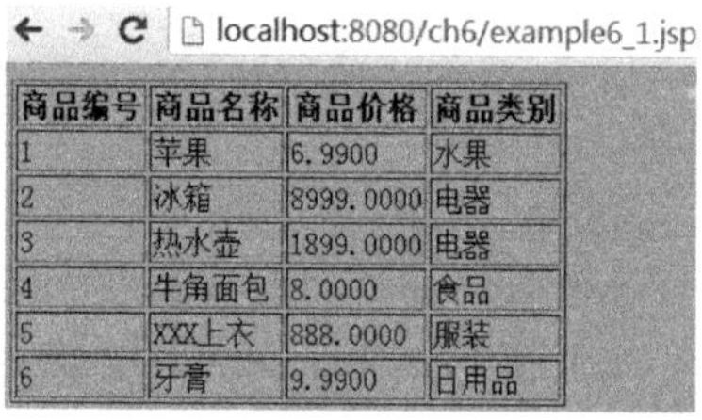

商品编号	商品名称	商品价格	商品类别
1	苹果	6.9900	水果
2	冰箱	8999.0000	电器
3	热水壶	1899.0000	电器
4	牛角面包	8.0000	食品
5	XXX上衣	888.0000	服装
6	牙膏	9.9900	日用品

图 6.8 使用 JDBC-ODBC 桥接器连接数据库

例 6.1 页面文件 example6_1. jsp 的代码如下：

```
<%@ page language = "java" contentType = "text/html; charset = GBK" pageEncoding = "GBK" %>
<%@ page import = "java.sql.*" %>
<html>
    <head>
        <title>example6_1.jsp</title>
    </head>
    <body bgcolor = "lightgreen">
        <%
            Connection con = null;
            Statement st = null;
            ResultSet rs = null;
            try {
                Class.forName("sun.jdbc.odbc.JdbcOdbcDriver");
            } catch (ClassNotFoundException e) {
```

```
                    e.printStackTrace();
                }
                try {
                    con = DriverManager.getConnection("jdbc:odbc:myGod","firstDB", "firstDB");
                    st = con.createStatement();
                    rs = st.executeQuery("select * from goodsInfo");
                    out.print("<table border = 1>");
                    out.print("<tr>");
                        out.print("<th>商品编号</th>");
                        out.print("<th>商品名称</th>");
                        out.print("<th>商品价格</th>");
                        out.print("<th>商品类别</th>");
                    out.print("</tr>");
                    while(rs.next()){
                        out.print("<tr>");
                            out.print("<td>" + rs.getString(1) + "</td>");
                            out.print("<td>" + rs.getString(2) + "</td>");
                            out.print("<td>" + rs.getString(3) + "</td>");
                            out.print("<td>" + rs.getString(4) + "</td>");
                        out.print("</tr>");
                    }
                    out.print("</table>");
                } catch (SQLException e) {
                    e.printStackTrace();
                } finally{
                    try{
                        if(rs!= null){
                            rs.close();
                        }
                        if(st!= null){
                            st.close();
                        }
                        if(con!= null){
                            con.close();
                        }
                    }catch (SQLException e) {
                        e.printStackTrace();
                    }
                }
            %>
    </body>
</html>
```

注意：运行例6.1的页面example6_1.jsp时，可能会出现"[Microsoft][ODBC驱动程序管理器]在指定的DSN中,驱动程序和应用程序之间的体系结构不匹配"异常。产生该异常的原因是JDK版本与Microsoft Office不匹配。版本要求是,64位Microsoft Office使用64位JDK连接Access；32位Microsoft Office使用32位JDK连接Access。

6.1.4 实践环节

（1）参考本节任务中的主要内容,创建数据源mySky,该数据源指定的数据库是

goods.accdb。

（2）编写一个JSP页面practice6_1.jsp，该页面中的Java程序片代码使用JDBC-ODBC桥接器连接到数据源mySky，查询goodsInfo表中goodsPrice字段值大于100的全部记录。

6.2 使用纯Java数据库驱动程序连接数据库

6.2.1 核心知识

使用纯Java数据库驱动程序连接数据库，需要以下两个步骤。

（1）注册纯Java数据库驱动程序。

（2）和指定的数据库建立连接。

下面以Oracle 10g为例，讲解如何使用纯Java数据库驱动程序连接数据库。

1. 注册纯Java数据库驱动程序

每种数据库都配有自己的纯Java数据库驱动程序。Oracle 10g的纯Java驱动程序一般位于数据库安装目录\oracle\product\10.2.0\db_1\jdbc\lib下，名为classes12.jar。

为了连接Oracle 10g数据库，可以将classes12.jar文件复制到Web应用程序的/WEB-INF/lib目录下。然后，通过java.lang.Class类的forName()，动态注册Oracle 10g的纯Java驱动程序，代码如下：

```
try {
    Class.forName("oracle.jdbc.driver.OracleDriver");
} catch (ClassNotFoundException e) {
    e.printStackTrace();
}
```

2. 和指定的数据库建立连接

和Oracle数据库建立连接的代码如下：

```
try {
    Connection con = DriverManager.getConnection("jdbc:oracle:thin:@主机:端口号:数据库名","用户名","密码");
} catch (SQLException e) {
    e.printStackTrace();
}
```

其中，主机是安装Oracle服务器的IP地址，如果是本机则为localhost；Oracle默认端口号为1521；Oracle默认数据库名为orcl；用户名和密码是访问Oracle服务器的用户权限。

应用程序连接Oracle数据库时，必须事先启动Oracle服务器的OracleServiceORCL和OracleOraDb10g_home1TNSListener两个服务，否则会抛出连接异常。

6.2.2 能力目标

掌握纯Java数据库驱动程序连接数据库的方法。

6.2.3　任务驱动

1. 任务的主要内容

编写一个 JSP 页面 example6_2.jsp，该页面中的 Java 程序片代码使用纯 Java 驱动程序连接 Oracle 数据库，查询 goodsInfo 表中的全部记录。创建 goodsInfo 表的 SQL 语句如下：

```
create table goodsinfo (
    goodsId number(4) not null,
    goodsName varchar(50) not null,
    goodsPrice number(7,2) not null,
    goodsType varchar(10) not null,
    constraint pk_goodsinfo primary key (goodsId)
);
insert into goodsinfo values(1,'牙膏',12,'日用品');
insert into goodsinfo values(2,'冰箱',2500,'电器');
insert into goodsinfo values(3,'蛋糕',28,'食品');
insert into goodsinfo values(4,'苹果',48,'水果');
insert into goodsinfo values(5,'上衣',1800,'服装');
insert into goodsinfo values(6,'书包',78,'文具');
commit;
```

页面运行效果如图 6.9 所示。

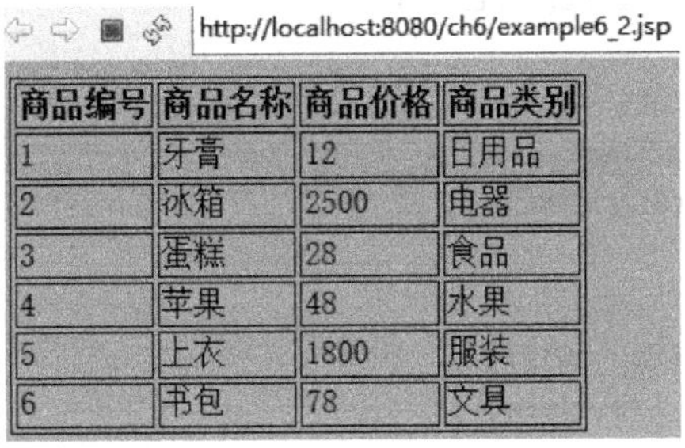

商品编号	商品名称	商品价格	商品类别
1	牙膏	12	日用品
2	冰箱	2500	电器
3	蛋糕	28	食品
4	苹果	48	水果
5	上衣	1800	服装
6	书包	78	文具

图 6.9　使用纯 Java 驱动程序连接 Oracle 数据库

2. 任务的代码模板

example6_2.jsp 的代码模板如下：

```
<%@ page language = "java" contentType = "text/html; charset = GBK"
    pageEncoding = "GBK" %>
<%@ page import = "java.sql.*" %>
<html>
    <head>
        <title>example5_2.jsp</title>
    </head>
    <body bgcolor = "LightPink">
        <%
            Connection con = null;
            Statement st = null;
            ResultSet rs = null;
```

```
        try {
            【代码1】//注册Oracle的纯Java驱动程序
        } catch (ClassNotFoundException e) {
            e.printStackTrace();
        }
        try {
            con = DriverManager.getConnection("jdbc:oracle:thin:@localhost:1521:orcl",
            "system","system");
            st = con.createStatement();
            rs = st.executeQuery("select * from goodsInfo");
            out.print("<table border=1>");
            out.print("<tr>");
                out.print("<th>商品编号</th>");
                out.print("<th>商品名称</th>");
                out.print("<th>商品价格</th>");
                out.print("<th>商品类别</th>");
            out.print("</tr>");
            while(rs.next()){
                out.print("<tr>");
                    out.print("<td>" + rs.getString(1) + "</td>");
                    out.print("<td>" + rs.getString(2) + "</td>");
                    out.print("<td>" + rs.getString(3) + "</td>");
                    out.print("<td>" + rs.getString(4) + "</td>");
                out.print("</tr>");
            }
            out.print("</table>");
        } catch (SQLException e) {
            e.printStackTrace();
        } finally{
            try{
                if(rs!=null){
                    rs.close();
                }
                if(st!=null){
                    st.close();
                }
                if(con!=null){
                    con.close();
                }
            }catch (SQLException e) {
                e.printStackTrace();
            }
        }
    %>
    </body>
</html>
```

3. 任务小结或知识扩展

从任务中可以看出编写程序访问数据库需要有以下几个步骤。

1）导入 java.sql 包

所有与数据库有关的对象和方法都在 java.sql 包中，包 java.sql 包含用 Java 操作关系数据库的类和接口。因此在使用 JDBC 的程序中必须要加入 import java.sql.*。

2）加载驱动程序

在该任务中使用了 Class 类（java.lang 包）中的方法 forName，来装入该驱动程序的类定义 oracle.jdbc.driver.OracleDriver，从而创建了该驱动程序的一个实例。

3）连接数据库

完成上述操作后，就可以连接一个特定的数据库了。这需要创建 Connection 类的一个实例，并使用 DriverManager 的方法 getConnection 来尝试建立用 URL 指定的数据库的连接。代码如下：

```
con = DriverManager.getConnection("jdbc:oracle:thin:@localhost:1521:orcl","system",
"system");
```

4）访问数据库

访问数据库，需要先用 Connection 类的 createStatement 方法从指定的数据库连接得到一个 Statement 的实例，然后用这个实例的 executeQuery 方法来执行一条 SQL 语句。代码如下：

```
st = con.createStatement();
rs = st.executeQuery("select * from goodsInfo");
```

5）处理返回的结果集

ResultSet 对象是 JDBC 中比较重要的一个对象，几乎所有的查询操作都将数据作为 ResultSet 对象返回。处理结果集 ResultSet 对象的代码如下：

```
while(rs.next()){
    out.print("<tr>");
    out.print("<td>" + rs.getString(1) + "</td>");
    out.print("<td>" + rs.getString(2) + "</td>");
    out.print("<td>" + rs.getString(3) + "</td>");
    out.print("<td>" + rs.getString(4) + "</td>");
    out.print("</tr>");
}
```

6）关闭数据库连接，释放资源

对数据库的操作完成之后，要及时关闭 ResultSet 对象、Statement 对象和数据库连接对象 Connection，从而释放占用的资源，这就要用到 close 方法。代码如下：

```
rs.close();
st.close();
con.close();
```

关闭的顺序从先到后依次为 ResultSet 对象、Statement 对象和 Connection 对象。

4. 代码模板的参考答案

【代码 1】：`Class.forName("oracle.jdbc.driver.OracleDriver");`

6.2.4 实践环节

编写一个 JSP 页面 practice6_2.jsp，该页面中的 Java 程序片代码使用纯 Java 驱动程序连接 Oracle 数据库，查询 goodsInfo 表中 goodsPrice 字段值大于 10 并小于 50 的全部记录。页面运行效果如图 6.10 所示。

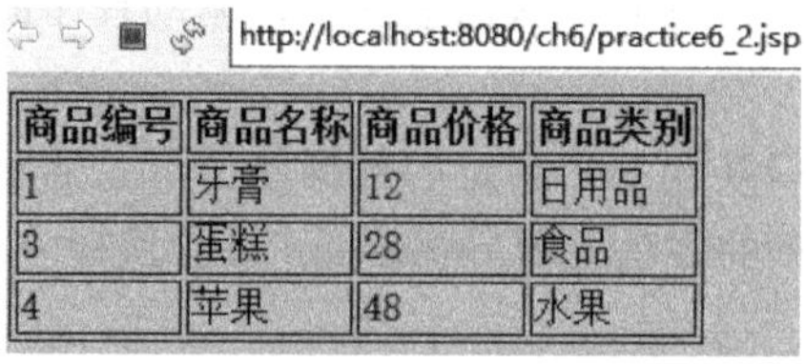

商品编号	商品名称	商品价格	商品类别
1	牙膏	12	日用品
3	蛋糕	28	食品
4	苹果	48	水果

图 6.10　practice6_2.jsp 页面运行效果

6.3　Statement、ResultSet 的使用

6.3.1 核心知识

和数据库建立连接后，接下来要执行 SQL 语句，需要有以下几个步骤。

1. 创建 Statement 对象

Statement 对象代表一条发送到数据库执行的 SQL 语句。由已创建的 Connection 对象 con 调用 createStatement()方法来创建 Statement 对象，代码如下：

```
Statement smt = con.createStatement();
```

2. 执行 SQL 语句

创建 Statement 对象后，可以使用 Statement 对象调用 executeUpdate(String sql)、executeQuery(String sql)等方法来执行 SQL 语句。

executeUpdate(String sql)方法主要用于执行 INSERT、UPDATE 或 DELETE 语句以及 SQL DDL 语句，例如 CREATE TABLE 和 DROP TABLE。该方法返回一个整数(代表被更新的行数)，对于 CREATE TABLE 和 DROP TABLE 等不操作行的指令，返回零。

executeQuery(String sql)方法则是用于执行 SELECT 等查询数据库的 SQL 语句，该方法返回 ResultSet 对象，代表查询的结果。

3. 处理返回的 ResultSet 对象

ResultSet 对象是 executeQuery(String sql)方法的返回值，被称为结果集，它代表符合 SQL 语句条件的所有行。ResultSet 对象调用 next()方法移动到下一个数据行(顺序查询)，当数据行存在时，next()方法返回 true，否则返回 false。获得一行数据后，ResultSet 对象可以使用 getXxx 方法获得字段值，getXxx 方法都提供依字段名称取得数据，或依字段顺序取得数据的方法。

6.3.2 能力目标

能够灵活使用 Statement 与 ResultSet 对象对数据库进行增删改查。

6.3.3 任务驱动

1. 任务的主要内容

编写两个 JSP 页面：addGoods.jsp 和 showAllGoods.jsp。用户可以在 addGoods.jsp 页面中输入信息后，单击"添加"按钮把信息添加到 goodsInfo 表中。然后，在 showAllGoods.jsp 页面中显示所有商品信息。在该任务中需要编写一个 bean(GoodsBean.java)用来添加和查询记录。页面运行效果如图 6.11 所示。

(a) 添加记录

goodsId	goodsName	goodsPrice	goodsType
1	牙膏	12	日用品
2	冰箱	2500	电器
3	蛋糕	28	食品
4	苹果	48	水果
5	上衣	1800	服装
6	书包	78	文具
7	香蕉	17	水果

(b) 查询记录

图 6.11　页面运行效果

2. 任务的代码模板

addGoods.jsp 的代码如下：

```jsp
<%@ page language = "java" contentType = "text/html; charset = GBK" pageEncoding = "GBK" %>
<html>
    <head>
        <title>addGoods.jsp</title>
    </head>
    <body>
        <h4>商品编号是主键,不能重复,每个信息都必须输入!</h4>
        <form action = "showAllGoods.jsp" method = "post">
        <table border = "1">
            <tr>
                <td>商品编号:</td>
                <td><input type = "text" name = "goodsId"/></td>
            </tr>
            <tr>
                <td>商品名称:</td>
```

```
            <td><input type = "text" name = "goodsName"/></td>
        </tr>
        <tr>
            <td>商品价格:</td>
            <td><input type = "text" name = "goodsPrice"/></td>
        </tr>
        <tr>
            <td>商品类型:</td>
            <td>
                <select name = "goodsType">
                    <option value = "日用品">日用品
                    <option value = "电器">电器
                    <option value = "食品">食品
                    <option value = "水果">水果
                    <option value = "服装">服装
                    <option value = "文具">文具
                    <option value = "其他">其他
                </select>
            </td>
        </tr>
        <tr>
            <td><input type = "submit" value = "添加"></td>
            <td><input type = "reset" value = "重置"></td>
        </tr>
    </table>
    </form>
</body>
</html>
```

showAllGoods.jsp 的代码如下:

```
<%@ page language = "java" contentType = "text/html; charset = GBK" pageEncoding = "GBK" %>
<%@ page import = "bean.GoodsBean" %>
<html>
    <head>
        <title>showAllGoods.jsp</title>
    </head>
    <body>
        <%
            request.setCharacterEncoding("GBK");
        %>
        <jsp:useBean id = "goods" class = "bean.GoodsBean" scope = "page"></jsp:useBean>
        <jsp:setProperty property = "*" name = "goods"/>
        <%
            goods.addGoods();           //添加商品
        %>
        <jsp:getProperty property = "queryResult" name = "goods"/><!-- 获得查询结果 -->
    </body>
</html>
```

GoodsBean.java 的代码模板如下:

```java
package bean;
import java.sql.*;
public class GoodsBean {
    int goodsId;
    String goodsName;
    double goodsPrice;
    String goodsType;
    StringBuffer queryResult;                        //查询结果
    public GoodsBean(){
    }
    public int getGoodsId() {
        return goodsId;
    }
    public void setGoodsId(int goodsId) {
        this.goodsId = goodsId;
    }
    public String getGoodsName() {
        return goodsName;
    }
    public void setGoodsName(String goodsName) {
        this.goodsName = goodsName;
    }
    public double getGoodsPrice() {
        return goodsPrice;
    }
    public void setGoodsPrice(double goodsPrice) {
        this.goodsPrice = goodsPrice;
    }
    public String getGoodsType() {
        return goodsType;
    }
    public void setGoodsType(String goodsType) {
        this.goodsType = goodsType;
    }
    //添加记录
    public void addGoods(){
        Connection con = null;
        Statement st = null;
        try {
            Class.forName("oracle.jdbc.driver.OracleDriver");
        } catch (ClassNotFoundException e) {
            e.printStackTrace();
        }
        try {
            con = DriverManager.getConnection("jdbc:oracle:thin:@localhost:1521:orcl",
            "system","system");
            【代码1】                                   //创建 Statement 对象
            String addSql = "insert into goodsInfo values(" + goodsId + ",'" + goodsName + "','" + goodsPrice
            + ",'" + goodsType + "')";
            【代码2】                                   //执行 insert 语句
        }catch (SQLException e) {
```

```java
                e.printStackTrace();
            }finally{
                try{
                    if(st!=null){
                        st.close();
                    }
                    if(con!=null){
                        con.close();
                    }
                }catch (SQLException e) {
                    e.printStackTrace();
                }
            }
        }
        //查询记录
        public StringBuffer getQueryResult(){
            queryResult = new StringBuffer();
            Connection con = null;
            Statement st = null;
            ResultSet rs = null;
            try {
                Class.forName("oracle.jdbc.driver.OracleDriver");
            } catch (ClassNotFoundException e) {
                e.printStackTrace();
            }
            try {
                con = DriverManager.getConnection("jdbc:oracle:thin:@localhost:1521:orcl",
                    "system","system");
                【代码3】                                  //创建Statement对象
                String selectSql = "select * from goodsInfo";
                【代码4】                                  //执行select语句
                queryResult.append("<table border=1>");
                    queryResult.append("<tr>");
                        queryResult.append("<th>goodsId</th>");
                        queryResult.append("<th>goodsName</th>");
                        queryResult.append("<th>goodsPrice</th>");
                        queryResult.append("<th>goodsType</th>");
                    queryResult.append("</tr>");
                while(rs.next()){
                    queryResult.append("<tr>");
                        queryResult.append("<td>" + rs.getString(1) + "</td>");
                        queryResult.append("<td>" + rs.getString(2) + "</td>");
                        queryResult.append("<td>" + rs.getString(3) + "</td>");
                        queryResult.append("<td>" + rs.getString(4) + "</td>");
                    queryResult.append("</tr>");
                }
                queryResult.append("</table>");
            }catch (SQLException e) {
                e.printStackTrace();
            }finally{
                try{
```

```
            if(rs!= null){
                rs.close();
            }
            if(st!= null){
                st.close();
            }
            if(con!= null){
                con.close();
            }
        }catch (SQLException e) {
            e.printStackTrace();
        }
    }
    return queryResult;
    }
}
```

3. 任务小结或知识扩展

ResultSet 对象自动维护指向其当前数据行的游标。每调用一次 next()方法，游标向下移动一行。最初它位于结果集的第一行之前，因此第一次调用 next()，将把游标置于第一行上，使它成为当前行。随着每次调用 next()，导致游标向下移动一行，按照从上至下次序获取 ResultSet 行，实现顺序查询。

ResultSet 对象包含 SQL 语句的执行结果。它通过一套 get 方法对这些行的数据进行访问，即使用 getXxx()方法获得数据。get 方法很多，究竟用哪一个 getXxx()方法，由列的数据类型来决定。使用 getXxx()方法时，需要注意以下两点。

（1）无论列是何种数据类型，总可以使用 getString(int columnIndex)或 getString(String columnName)方法获得列值的字符串表示。

（2）如果使用 getString(int columnIndex)方法查看一行记录，不允许颠倒顺序，例如，不允许

```
rs.getString(2);
rs.getString(1);
```

4. 代码模板的参考答案

【代码 1】: st = con.createStatement();
【代码 2】: st.executeUpdate(addSql);
【代码 3】: st = con.createStatement();
【代码 4】: rs = st.executeQuery(selectSql);

6.3.4 实践环节

编写两个 JSP 页面：inputQuery.jsp 和 showGoods.jsp。用户可以在 inputQuery.jsp 页面输入查询条件后，单击"查询"按钮。然后，在 showGoods.jsp 页面中显示符合查询条件的商品信息。在本节任务的 bean(GoodsBean.java)中添加一个方法 getQueryResultBy()实现该题的条件查询功能。页面运行效果如图 6.12 所示。

(a) 输入条件

(b) 符合查询条件的记录

图 6.12　页面运行效果

6.4　游动查询

6.4.1　核心知识

有时候需要结果集的游标前后移动，这时可使用滚动结果集。为了获得滚动结果集，必须首先用下面的方法得到一个 Statement 对象：

Statement st = con.createStatement(int type, int concurrency);

根据 type 和 concurrency 的取值，当执行 ResultSet rs = st.executeQuery(String sql)时，会返回不同类型的结果集。

type 的取值决定滚动方式，它的取值如下。

(1) ResultSet.TYPE_FORWORD_ONLY：表示结果集只能向下滚动。

(2) ResultSet.TYPE_SCROLL_INSENSITIVE：表示结果集可以上下滚动，当数据库变化时，结果集不变。

(3) ResultSet.TYPE_SCROLL_SENSITIVE：表示结果集可以上下滚动，当数据库变化时，结果集同步改变。

concurrency 取值表示是否可以用结果集更新数据库，它的取值如下。

(1) ResultSet.CONCUR_READ_ONLY：表示不能用结果集更新数据库表。

(2) ResultSet.CONCUR_UPDATETABLE：表示能用结果集更新数据库表。

6.4.2　能力目标

能够灵活使用滚动结果集进行游动查询。

6.4.3　任务驱动

1. 任务的主要内容

编写一个 JSP 页面 randomQuery.jsp，查询 goodsInfo 表中的全部记录，并将结果逆序输出，最后单独输出第 4 条记录。运行结果如图 6.13 所示。

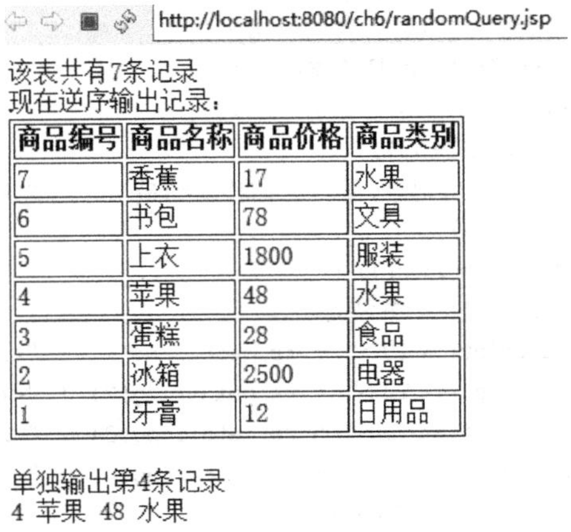
http://localhost:8080/ch6/randomQuery.jsp

该表共有7条记录
现在逆序输出记录：

商品编号	商品名称	商品价格	商品类别
7	香蕉	17	水果
6	书包	78	文具
5	上衣	1800	服装
4	苹果	48	水果
3	蛋糕	28	食品
2	冰箱	2500	电器
1	牙膏	12	日用品

单独输出第4条记录
4 苹果 48 水果

图 6.13　随机查询记录

2. 任务的代码模板

randomQuery.jsp 的代码模板如下：

```
<%@ page language = "java" contentType = "text/html; charset = GBK" pageEncoding = "GBK" %>
<%@ page import = "java.sql.*" %>
<html>
    <head>
        <title>randomQuery.jsp</title>
    </head>
    <body>
        <%
            Connection con = null;
            Statement st = null;
            ResultSet rs = null;
            try {
                Class.forName("oracle.jdbc.driver.OracleDriver");
            } catch (ClassNotFoundException e) {
                e.printStackTrace();
            }
            try {
                con = DriverManager.getConnection("jdbc:oracle:thin:@localhost:1521:orcl",
                "system","system");
                //创建 st 对象,该对象获得的结果集和数据库同步变化,但不能用结果集更新表
                【代码 1】
                //返回可滚动的结果集
                rs = st.executeQuery("SELECT * FROM goodsInfo");
                //将游标移动到最后一行
                rs.last();
                //获取最后一行的行号
                int lownumber = rs.getRow();
                out.print("该表共有" + lownumber + "条记录");
                out.print("<BR>现在逆序输出记录: ");
                out.print("<Table Border = 1>");
```

```
                    out.print("<TR>");
                    out.print("<TH>商品编号</TH>");
                    out.print("<TH>商品名称</TH>");
                    out.print("<TH>商品价格</TH>");
                    out.print("<TH>商品类别</TH>");
                    out.print("</TR>");
                    //为了逆序输出记录,需将游标移动到最后一行之后
                    rs.afterLast();
                    while (rs.previous()) {
                        out.print("<TR>");
                        out.print("<TD>" + rs.getString(1) + "</TD>");
                        out.print("<TD>" + rs.getString(2) + "</TD>");
                        out.print("<TD>" + rs.getString(3) + "</TD>");
                        out.print("<TD>" + rs.getString(4) + "</TD>");
                        out.print("</TR>");
                    }
                    out.print("</Table>");
                    out.print("<BR>单独输出第 4 条记录<BR>");
                    rs.absolute(4);
                    out.print(rs.getString(1) + " ");
                    out.print(rs.getString(2) + " ");
                    out.print(rs.getString(3) + " ");
                    out.print(rs.getString(4));
                } catch (SQLException e) {
                    e.printStackTrace();
                } finally {
                    try {
                        if (rs != null) {
                            rs.close();
                        }
                        if (st != null) {
                            st.close();
                        }
                        if (con != null) {
                            con.close();
                        }
                    } catch (SQLException e) {
                        e.printStackTrace();
                    }
                }
            %>
        </body>
</html>
```

3. 任务小结或知识扩展

游动查询经常用到 ResultSet 类的下述方法。

(1) public boolean absolute(int row): 将游标移到参数 row 指定的行。如果 row 取负值,就是倒数的行,如-1 表示最后一行。当移到最后一行之后或第一行之前时,该方法返回 false。

（2）public voidafterLast()：将游标移到结果集的最后一行之后。

（3）public void beforeFirst()：将游标移到结果集的第一行之前。

（4）public void first()：将游标移到结果集的第一行。

（5）public int getRow()：得到当前游标所指定的行号，如果没有行，则返回 0。

（6）public boolean is AfterLast()：判断游标是不是在结果集的最后一行之后。

（7）public boolean is BeforeFirst()：判断游标是不是在结果集的第一行之前。

（8）public void last()：将游标移到结果集的最后一行。

（9）public boolean previous()：将游标向上移动（和 next 方法相反），当移到结果集的第一行之前时返回 false。

4. 代码模板的参考答案

【代码 1】：st = con.createStatement(ResultSet.TYPE_SCROLL_SENSITIVE,
　　　　　　ResultSet.CONCUR_READ_ONLY);

6.4.4 实践环节

编写一个 JSP 页面 practice6_4.jsp，查询 goodsInfo 表中的记录，并逆序输出偶数行的记录。运行结果如图 6.14 所示。

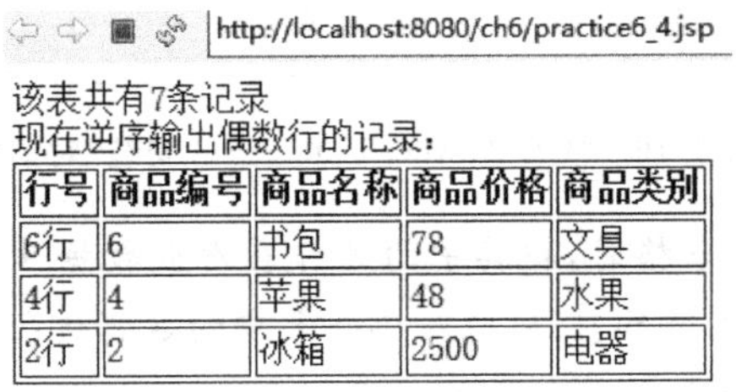

图 6.14　逆序输出偶数行的记录

6.5　访问 Excel 电子表格

6.5.1 核心知识

可以使用 JDBC-ODBC 桥接器访问 Excel 电子表格，包括增加、删除、修改、查询等操作。但访问 Excel 电子表格和访问关系数据库有所不同。访问 Excel 电子表格要经过以下两个步骤。

1. 创建 Excel 电子表格

使用 Excel 2007 创建一个名为 student.xlsx 的电子表格，该 Excel 表格中有一个名为 studentScore 的工作表。student.xlsx 电子表格如图 6.15 所示。

2. 创建数据源

创建 Excel 电子表格后，需要为它创建数据源。数据源的名字是 student，为数据源选择的驱动程序是 Microsoft Excel Driver（*.xls，*.xlsx，*.xlsm，*.xlsb）。如果想通

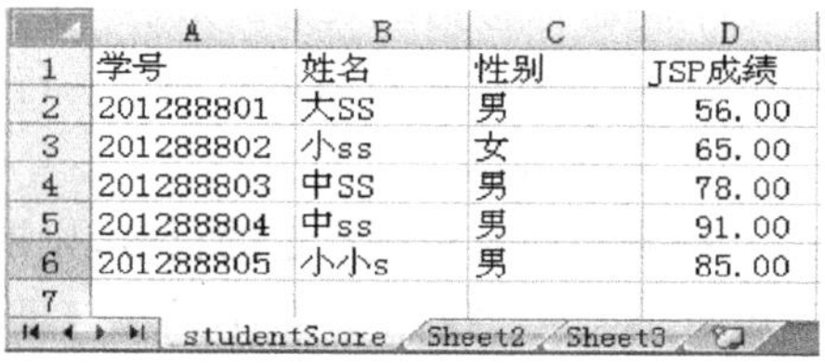

	A	B	C	D
1	学号	姓名	性别	JSP成绩
2	201288801	大SS	男	56.00
3	201288802	小ss	女	65.00
4	201288803	中SS	男	78.00
5	201288804	中ss	男	91.00
6	201288805	小小s	男	85.00
7				

图 6.15　student.xlsx 电子表格

过 JDBC-ODBC 修改 Excel 电子表格，在设置数据源时，单击“选项”按钮，将“只读”属性设置为未选中状态，如图 6.16 所示。

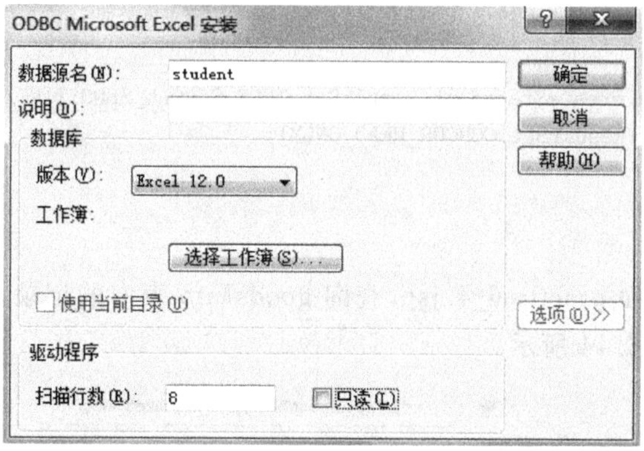

图 6.16　“ODBC Microsofe Excel 安装”对话框

注意：*访问 Excel 电子表格时，把其中的工作表看成数据库中的表，如 student.xlsx 中的 studentScore 工作表。使用 SQL 语句对工作表中的数据进行增加、删除、修改和查询时，要在表名前加[，在表名后加$]，如 select * from [studentScore$]。*

6.5.2　能力目标

能够灵活使用 JDBC-ODBC 桥接器访问 Excel 电子表格。

6.5.3　任务驱动

1. 任务的主要内容

编写一个 JSP 页面 readExcel.jsp，在该页面的 Java 程序片中首先增加一条记录到 studentScore 工作表中，然后修改某条记录，最后查询全部记录。

2. 任务的代码模板

readExcel.jsp 的代码模板如下：

```
<%@ page language = "java" contentType = "text/html; charset = GBK" pageEncoding = "GBK" %>
<%@ page import = "java.sql.*" %>
<html>
    <head>
        <title>readExcel.jsp</title>
    </head>
```

```jsp
<body bgcolor = "lightgreen">
<%
    Connection con = null;
    Statement st = null;
    ResultSet rs = null;
    try {
        Class.forName("sun.jdbc.odbc.JdbcOdbcDriver");
    } catch (ClassNotFoundException e) {
        e.printStackTrace();
    }
    try {
        con = DriverManager.getConnection("jdbc:odbc:student","", "");
        st = con.createStatement();
        //添加记录
        st.executeUpdate("insert into 【代码1】 values('201212009','sss','nv',73)");
        //修改记录
        st.executeUpdate("update 【代码2】 set 性别 = '女' where 学号 = '201288803'");
        //查询全部记录
        rs = st.executeQuery("select * from 【代码3】 ");
        out.print("<table border = 1>");
        out.print("<tr>");
            out.print("<th>学号</th>");
            out.print("<th>姓名</th>");
            out.print("<th>性别</th>");
            out.print("<th>JSP 成绩</th>");
        out.print("</tr>");
        while(rs.next()){
            out.print("<tr>");
                out.print("<td>" + rs.getString(1) + "</td>");
                out.print("<td>" + rs.getString(2) + "</td>");
                out.print("<td>" + rs.getString(3) + "</td>");
                out.print("<td>" + rs.getString(4) + "</td>");
            out.print("</tr>");
        }
        out.print("</table>");
    } catch (SQLException e) {
        e.printStackTrace();
    }finally{
        try{
            if(rs!= null){
                rs.close();
            }
            if(st!= null){
                st.close();
            }
            if(con!= null){
                con.close();
            }
        }catch (SQLException e) {
            e.printStackTrace();
        }
```

```
            }
        %>
    </body>
</html>
```

3. 任务小结或知识扩展

一个 Excel 电子表格可以有多个工作表，使用 JDBC-ODBC 可以访问电子表格中任何一个工作表，就像访问一个数据库中任意一张表一样。

注意：访问 Excel 与访问 Access 一样，也可能会出现"[Microsoft][ODBC 驱动程序管理器]在指定的 DSN 中，驱动程序和应用程序之间的体系结构不匹配"异常，解决方法与 6.1.1 节中的 Access 一样。

4. 代码模板的参考答案

【代码 1】：[studentScore $]
【代码 2】：[studentScore $]
【代码 3】：[studentScore $]

6.5.4 实践环节

在 student.xlsx 电子表格中新建一个工作表 empTable，如图 6.17 所示，编写一个 JSP 页面 practice6_5.jsp，在该页面中显示 empTable 表中的所有记录。

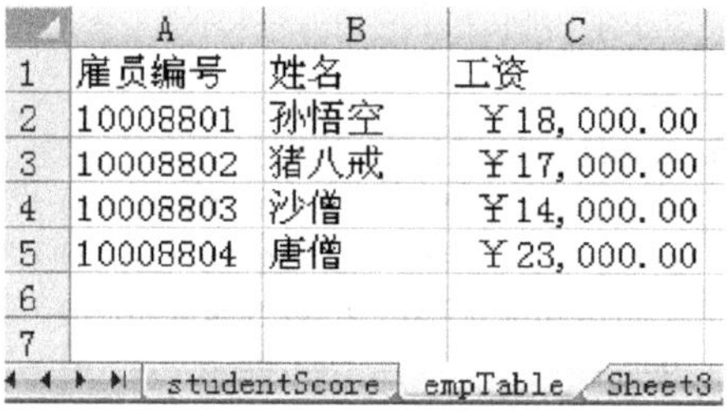

	A	B	C
1	雇员编号	姓名	工资
2	10008801	孙悟空	￥18,000.00
3	10008802	猪八戒	￥17,000.00
4	10008803	沙僧	￥14,000.00
5	10008804	唐僧	￥23,000.00
6			
7			

图 6.17　工作表 empTable

6.6　使用连接池

6.6.1 核心知识

与数据库建立连接是一个耗资源的活动，每次都得花费一定的时间。这个时间对于一次或几次数据库连接，系统的开销或许不明显。可是对于同时有成千上万人频繁地进行数据库连接操作的大型电子商务网站来说势必占用很多系统资源，网站的响应速度必定下降，严重时甚至会造成服务器的崩溃。因此，合理地建立数据库连接是非常重要的。

数据库连接池的基本思想是，为数据库连接建立一个"缓冲池"。预先在"缓冲池"中放入一定数量的连接，当需要建立数据库连接时，只需从"缓冲池"中取出一个，使用完毕之后再放回去。可以通过设定连接池最大连接数来防止系统无限度地与数据库连接。更为重要的是，通过连接池的管理机制监视数据库连接的数量及使用情况，为系统开发、测试和性能调整提供依据。其工作原理如图 6.18 所示。

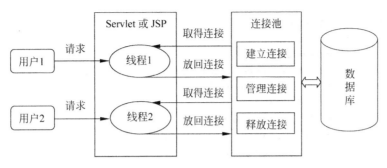

图 6.18　连接池的原理

6.6.2 能力目标

了解连接池的工作原理,灵活使用连接池连接数据库。

6.6.3 任务驱动

1. 任务的主要内容

编写一个 JSP 页面 conPool.jsp,在该页面中使用 scope 为 application 的 bean(由 ConnectionPool 类负责创建)。该 bean 创建时,将建立一定数量的连接对象。因此,所有用户将共享这些连接对象。在 JSP 页面中使用 bean 获得一个连接对象,然后使用该连接对象访问数据库中的 goodsInfo 表(查询出商品价格大于 500 的商品)。页面运行效果如图 6.19 所示。

商品编号	商品名称	商品价格	商品类别
2	冰箱	2500	电器
5	上衣	1800	服装

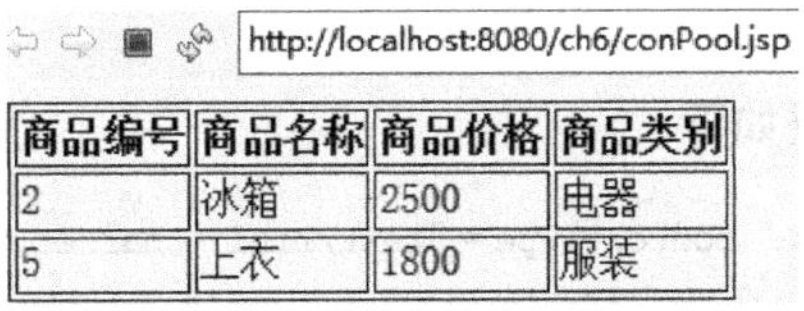

图 6.19　使用连接池连接数据库

2. 任务的代码模板

ConnectionPool.java 的代码如下:

```
package db.connection.pool;
import java.sql.*;
import java.util.ArrayList;
public class ConnectionPool {
    //存放 Connection 对象的数组,数组被看成连接池
    ArrayList<Connection> list = new ArrayList<Connection>();
    //构造方法,创建 15 个连接对象,放到连接池中
    public ConnectionPool(){
        try {
            Class.forName("oracle.jdbc.driver.OracleDriver");
        } catch (ClassNotFoundException e) {
```

```java
                e.printStackTrace();
            }
            for(int i = 0;i < 15;i++){
                try {
                    Connection
                    con = DriverManager.getConnection("jdbc:oracle:thin:@localhost:1521:
                    orcl","system","system");
                    list.add(con);
                } catch (SQLException e) {
                    //TODO Auto-generated catch block
                    e.printStackTrace();
                }
            }
        }
        //从连接池中取出一个连接对象
        public synchronized Connection getOneCon(){
            if(list.size()> 0){
                //删除数组的第一个元素,并返回该元素中的连接对象
                return list.remove(0);
            }else{
                //连接对象被用完
                return null;
            }
        }
        //把连接对象放回连接池中
        public synchronized void releaseCon(Connection con){
            list.add(con);
        }
}
```

conPool.jsp 的代码模板如下:

```jsp
<%@ page language = "java" contentType = "text/html; charset = GBK" pageEncoding = "GBK" %>
<%@ page import = "java.sql.*" %>
<%@ page import = "db.connection.pool.*" %>
<html>
    <head>
        <title>使用连接池连接数据库</title>
    </head>
    <jsp:useBean id = "conpool" class = "db.connection.pool.ConnectionPool" scope = "application"/>
    <body bgcolor = "AliceBlue">
    <%
        Connection con = null;
        Statement st = null;
        ResultSet rs = null;
        try{
            【代码1】//使用 conpool 对象调用 getOneCon 方法从连接池中获得一个连接对象
            if(con == null){
                out.print("人数过多,稍后访问");
                return;
```

```
            }
            st = con.createStatement();
            rs = st.executeQuery("select * from goodsInfo where goodsPrice>500");
            out.print("<table border=1>");
                out.print("<tr>");
                    out.print("<th>商品编号</th>");
                    out.print("<th>商品名称</th>");
                    out.print("<th>商品价格</th>");
                    out.print("<th>商品类别</th>");
                out.print("</tr>");
            while(rs.next()){
                out.print("<tr>");
                    out.print("<td>" + rs.getString(1) + "</td>");
                    out.print("<td>" + rs.getString(2) + "</td>");
                    out.print("<td>" + rs.getString(3) + "</td>");
                    out.print("<td>" + rs.getString(4) + "</td>");
                out.print("</tr>");
            }
            out.print("</table>");
        }catch(SQLException e){
            e.printStackTrace();
        }finally{
            try{
                if(rs!=null){
                    rs.close();
                }
                if(st!=null){
                    st.close();
                }
                if(con!=null){
                    【代码2】//使用conpool对象调用releaseCon方法把连接对象放回连接池中
                }
            }catch(SQLException e){
                e.printStackTrace();
            }
        }
    %>
    </body>
</html>
```

3. 任务小结或知识扩展

第一次访问连接池时,需要耗费一定时间,这是因为在第一次访问时连接池中没有可用连接,但是在第二次访问时连接池中就有了一些可用的连接,可以直接从连接池中获得连接来访问数据库。

4. 代码模板的参考答案

【代码1】: con = conpool.getOneCon();
【代码2】: conpool.releaseCon(con);

6.6.4 实践环节

编写一个 JSP 页面 pratice6_6.jsp，在该页面中使用和任务中同样的 bean 获得一个数据库连接对象，然后使用该连接对象查询 goodsInfo 表中的全部记录。

6.7 其他典型数据库的连接

6.7.1 核心知识

使用 JDBC-ODBC 桥接器连接不同类型数据库的程序流程是类似的，只是在为 ODBC 数据源选择驱动程序时，选择对应的驱动程序即可，但有的数据库不被 ODBC 支持，如 MySQL 数据库。

使用纯 Java 数据库驱动程序连接不同类型数据库的程序流程和框架是基本相同的，需要重点关注的是，在连接各数据库时驱动程序加载部分的代码和连接部分的代码。下面分别介绍加载纯 Java 数据库驱动程序连接 SQL Server 2005 和 MySQL 5.5。这里假定要访问的数据库名为 mydatabase。

1. 连接 SQL Server 2005

1）获取纯 Java 数据库驱动程序

可以登录微软的官方网站 http://www.microsoft.com 下载 Microsoft SQL Server 2005 JDBC Driver 1.2，解压 Microsoft SQL Server 2005 jdbc driver1.2.exe 后得到 sqljdbc.jar 文件。然后，把 sqljdbc.jar 文件复制到 Web 应用程序的/WEB-INF/lib 文件夹中。

2）加载驱动程序

```
Class.forName("com.microsoft.sqlserver.jdbc.SQLServerDriver");
```

3）建立连接

```
Connection con =
    DriverManager.getConnection("jdbc:sqlserver://localhost:1433;DatabaseName=mydatabase","用户名","密码");
```

2. 连接 MySQL 5.5

1）获取纯 Java 数据库驱动程序

可以登录 MySQL 的官方网站 http://www.mysql.com 下载驱动程序。这里下载的是 mysql-connector-java-5.0.6-bin.jar。然后，把 mysql-connector-java-5.0.6-bin.jar 文件复制到 Web 应用程序的/WEB-INF/lib 文件夹中。

2）加载驱动程序

```
Class.forName("com.mysql.jdbc.Driver");
```

3）建立连接

```
Connection con =
    DriverManager.getConnection("jdbc:mysql://localhost:3306/mydatabase","用户名",
```

"密码");

6.7.2 能力目标

理解使用纯 Java 数据库驱动程序连接不同类型数据库的原理。

6.7.3 任务驱动

任务的主要内容如下。

(1) 安装 MySQL 5.5。
(2) 创建数据库和数据表。
(3) 连接数据库并操作数据表。

1. 安装 MySQL 5.5

登录官方网站下载 MySQL 5.5 安装程序 mysql-5.5.19-win32.msi，按照默认安装即可。用户名和密码都设置为 root。

2. 创建数据库和数据表

MySQL 5.5 安装后，选择"开始"/"程序"/MySQL/MySQL Server 5.5/MySQL 5.5 Command Line Client 命令。输入默认密码：root，成功启动 MySQL 监视器后，在 MS-DOS 窗口中出现 mysql>字样，如图 6.20 所示。

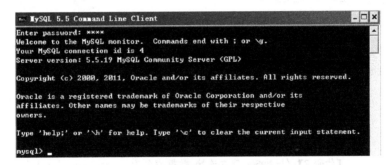

图 6.20　启动 MySQL 监视器

成功启动 MySQL 监视器后，就可以使用如下命令：

create database mydatabase;

创建数据库 mydatabase，为了在 mydatabase 数据库中创建表，必须先进入该数据库，命令如下：

use mydatabase

进入数据库 mydatabase 的操作如图 6.21 所示。

进入数据库后，就可以创建表 employee。创建表的命令如下：

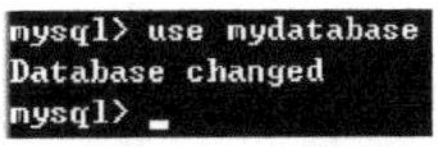

create table employee (
　　empNo varchar(10) not null,
　　name varchar(30) not null,
　　salary float not null,

图 6.21　进入数据库

```
    primary key (empNo)
);
insert into employee values('001','zhou',1200);
insert into employee values('002','wu',2500);
insert into employee values('003','zheng',2800);
insert into employee values('004','wang',4800);
insert into employee values('005','zhao',1800);
insert into employee values('006','qian',7800);
commit;
```

3. 连接数据库并操作数据表

编写一个JSP页面useMySQL.jsp,在该页面的Java程序片中连接MySQL数据库,并从employee表中查询出工资大于2000的雇员。

1) 任务的代码模板

useMySQL.jsp的代码模板如下:

```
<%@ page language="java" contentType="text/html; charset=GBK" pageEncoding="GBK"%>
<%@ page import="java.sql.*"%>
<html>
    <head>
        <title>useMySQL.jsp</title>
    </head>
    <body bgcolor="lightblue">
        <%
            Connection con = null;
            Statement st = null;
            ResultSet rs = null;
            try {
                【代码1】//加载MySQL的Java驱动
            } catch (ClassNotFoundException e) {
                e.printStackTrace();
            }
            try {
                【代码2】//与MySQL建立连接,数据库名为mydatabase,用户名和密码都为root
                st = con.createStatement();
                rs = st.executeQuery("select * from employee where salary>2000");
                out.print("<table border=1>");
                out.print("<tr>");
                    out.print("<th>雇员编号</th>");
                    out.print("<th>雇员姓名</th>");
                    out.print("<th>工资</th>");
                out.print("</tr>");
                while(rs.next()){
                    out.print("<tr>");
                        out.print("<td>"+rs.getString(1)+"</td>");
                        out.print("<td>"+rs.getString(2)+"</td>");
                        out.print("<td>"+rs.getString(3)+"</td>");
                    out.print("</tr>");
                }
                out.print("</table>");
            } catch (SQLException e) {
                e.printStackTrace();
```

```
            }finally{
                try{
                    if(rs!= null){
                        rs.close();
                    }
                    if(st!= null){
                        st.close();
                    }
                    if(con!= null){
                        con.close();
                    }
                }catch (SQLException e) {
                    e.printStackTrace();
                }
            }
        %>
    </body>
</html>
```

2) 代码模板的参考答案

【代码 1】: Class.forName("com.mysql.jdbc.Driver");
【代码 2】: con = DriverManager.getConnection("jdbc:mysql://localhost:3306/mydatabase",
          "root", "root");

6.7.4 实践环节

参考本节任务中的主要内容,使用 MySQL 创建一个数据库 yourdatabase,并在该数据库中创建一张表 student,并编写程序操作该表。

6.8 PreparedStatement 的使用

与 Statement 语句一样,PreparedStatement 语句同样可以完成向数据库发送 SQL 语句,获取数据库操作结果的功能。PreparedStatement 语句习惯地称为预处理语句。

Statement 对象在每次执行 SQL 语句时都将该语句传送给数据库,然后数据库解释器负责将 SQL 语句转换成内部命令,并执行该命令,完成相应的数据库操作。使用这种机制,每次向数据库发送一条 SQL 语句时,都要先转化成内部命令,如果不断地执行程序,就会加重解释器的负担,影响执行的速度。而使用 PreparedStatement 对象,将 SQL 语句传送给数据库进行预编译,以后需要执行同一条语句时就不再需要重新编译,直接执行就可以了,这样就大大提高了数据库的执行速度。

6.8.1 核心知识

可以使用 Connection 的对象 con 调用 prepareStatement(String sql)方法对参数 sql 指定的 SQL 语句进行预先编译,生成数据库的底层命令,并将该命令封装在 PreparedStatement 对象中。对于 SQL 语句中变动的部分,可以使用通配符"?"代替。例如:

```
PreparedStatement ps = con.prepareStatement("insert into goodsInfo values(?,?,?,?)");
```

然后使用对应的 setXxx(int parameterIndex, xxx value)方法设置"?"代表的值,其中参数 parameterIndex 用来表示 SQL 语句中从左到右的第 parameterIndex 个通配符号,value 代表该通配符所代表的具体值。例如:

```
ps.setInt(1,9);
ps.setString(2, "手机");
ps.setDouble(3, 1900.8);
ps.setString(4, "通信");
```

若让 SQL 语句执行生效,需使用 PreparedStatement 的对象 ps 调用 executeUpdate()方法。如果是查询,ps 就调用 executeQuery()方法,并返回 ResultSet 对象。

6.8.2 能力目标

能够灵活使用预处理语句对象操作数据库。

6.8.3 任务驱动

1. 任务的主要内容

编写两个 JSP 页面:inputPrepareGoods.jsp 和 showPrepareGoods.jsp。用户可以在 inputPrepareGoods.jsp 页面中输入信息后,单击"添加"按钮把信息添加到 goodsInfo 表中。然后,在 showPrepareGoods.jsp 页面中显示所有商品信息。在该任务中需要编写一个 bean(UsePrepare.java),bean 中使用预处理语句向 goodsInfo 表中添加记录。页面运行效果如图 6.22 所示。

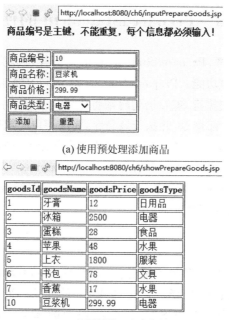

(a) 使用预处理添加商品

(b) 使用预处理查询商品

图 6.22 页面运行效果

2. 任务的代码模板

inputPrepareGoods.jsp 的代码如下：

```jsp
<%@ page language="java" contentType="text/html; charset=GBK" pageEncoding="GBK"%>
<html>
    <head>
        <title>使用预处理语句</title>
    </head>
    <body bgcolor="LightYellow">
        <h4>商品编号是主键,不能重复,每个信息都必须输入!</h4>
        <form action="showPrepareGoods.jsp" method="post">
        <table border="1">
            <tr>
                <td>商品编号:</td>
                <td><input type="text" name="goodsId"/></td>
            </tr>

            <tr>
                <td>商品名称:</td>
                <td><input type="text" name="goodsName"/></td>
            </tr>

            <tr>
                <td>商品价格:</td>
                <td><input type="text" name="goodsPrice"/></td>
            </tr>

            <tr>
                <td>商品类型:</td>
                <td>
                    <select name="goodsType">
                        <option value="日用品">日用品
                        <option value="电器">电器
                        <option value="食品">食品
                        <option value="水果">水果
                        <option value="服装">服装
                        <option value="文具">文具
                        <option value="水果">水果
                        <option value="电器">电器
                        <option value="其他">其他
                    </select>
                </td>
            </tr>

            <tr>
                <td><input type="submit" value="添加"></td>
                <td><input type="reset" value="重置"></td>
            </tr>
        </table>
        </form>
```

```
    </body>
</html>
```

showPrepareGoods.jsp 的代码如下：

```jsp
<%@ page language="java" contentType="text/html; charset=GBK" pageEncoding="GBK"%>
<%@ page import="bean.UsePrepare" %>
<html>
    <head>
        <title>使用预处理语句</title>
    </head>
    <body>
        <%
            request.setCharacterEncoding("GBK");
        %>
        <jsp:useBean id="prepareGoods" class="bean.UsePrepare" scope="page"></jsp:useBean>
        <jsp:setProperty property="*" name="prepareGoods"/>
        <%
            prepareGoods.addGoods();                    //添加商品
        %>
        <jsp:getProperty property="queryResult" name="prepareGoods"/><!-- 获得查询结果 -->
    </body>
</html>
```

UsePrepare.java 的代码模板如下：

```java
package bean;
import java.sql.*;
public class UsePrepare {
    int goodsId;
    String goodsName;
    double goodsPrice;
    String goodsType;
    StringBuffer queryResult;                    //查询所有商品结果
    public UsePrepare(){
    }
    public int getGoodsId() {
        return goodsId;
    }
    public void setGoodsId(int goodsId) {
        this.goodsId = goodsId;
    }
    public String getGoodsName() {
        return goodsName;
    }
    public void setGoodsName(String goodsName) {
        this.goodsName = goodsName;
    }
    public double getGoodsPrice() {
        return goodsPrice;
    }
    public void setGoodsPrice(double goodsPrice) {
```

```java
            this.goodsPrice = goodsPrice;
        }
        public String getGoodsType() {
            return goodsType;
        }
        public void setGoodsType(String goodsType) {
            this.goodsType = goodsType;
        }
//添加商品
public void addGoods(){
Connection con = null;
        PreparedStatement ps = null;
        try {
            Class.forName("oracle.jdbc.driver.OracleDriver");
        } catch (ClassNotFoundException e) {
            e.printStackTrace();
        }
        try {
            con = DriverManager.getConnection("jdbc:oracle:thin:@localhost:1521:orcl",
            "system","system");
            ps = con.prepareStatement("insert into goodsInfo values(?,?,?,?)");
            【代码1】//ps 调用 set 方法指定第 1 个通配符的值
            【代码2】//ps 调用 set 方法指定第 2 个通配符的值
            【代码3】//ps 调用 set 方法指定第 3 个通配符的值
            【代码4】//ps 调用 set 方法指定第 4 个通配符的值
            ps.executeUpdate();
        }catch (SQLException e) {
            e.printStackTrace();
        }finally{
            try{
                if(ps!= null){
                    ps.close();
                }
                if(con!= null){
                    con.close();
                }
            }catch (SQLException e) {
                e.printStackTrace();
            }
        }
    }
//获得所有商品信息
public StringBuffer getQueryResult(){
        queryResult = new StringBuffer();
        Connection con = null;
        PreparedStatement ps = null;
        ResultSet rs = null;
        try {
            Class.forName("oracle.jdbc.driver.OracleDriver");
        } catch (ClassNotFoundException e) {
            e.printStackTrace();
```

```
        }
        try {
            con = DriverManager.getConnection("jdbc:oracle:thin:@localhost:1521:orcl",
            "system","system");
            ps = con.prepareStatement("select * from goodsInfo");
            rs = ps.executeQuery();
            queryResult.append("<table border = 1>");
                queryResult.append("<tr>");
                    queryResult.append("<th>goodsId</th>");
                    queryResult.append("<th>goodsName</th>");
                    queryResult.append("<th>goodsPrice</th>");
                    queryResult.append("<th>goodsType</th>");
                queryResult.append("</tr>");
            while(rs.next()){
                queryResult.append("<tr>");
                    queryResult.append("<td>" + rs.getString(1) + "</td>");
                    queryResult.append("<td>" + rs.getString(2) + "</td>");
                    queryResult.append("<td>" + rs.getString(3) + "</td>");
                    queryResult.append("<td>" + rs.getString(4) + "</td>");
                queryResult.append("</tr>");
            }
            queryResult.append("</table>");
        }catch (SQLException e) {
            e.printStackTrace();
        }finally{
            try{
                if(rs!= null){
                    rs.close();
                }
                if(ps!= null){
                    ps.close();
                }
                if(con!= null){
                    con.close();
                }
            }catch (SQLException e) {
                e.printStackTrace();
            }
        }
        return queryResult;
    }
}
```

3. 任务小结或知识扩展

Statement 对象在执行 executeQuery(String sql)、executeUpdate(String sql)等方法时,如果 SQL 语句有些部分是动态的数据,必须使用"+"连字符组成完整的 SQL 语句,十分不便。例如,6.3.3 节任务中在添加商品时,必须按照如下方式组成 SQL 语句。

```
String addSql = "insert into goodsInfo values(" + goodsId + ",'" + goodsName + "'," +
goodsPrice + ",'" + goodsType + "')";
```

```
st.executeUpdate(addSql);
```

使用预处理语句不仅提高了数据库的访问效率,而且方便了程序的编写。预处理语句对象调用 executeUpdate() 和 executeQuery() 方法时不需要传递参数。例如:

```
int i = ps.executeUpdate();
```

或

```
ResultSet rs = ps.executeQuery();
```

4. 代码模板的参考答案

【代码1】: ps.setInt(1, goodsId);
【代码2】: ps.setString(2, goodsName);
【代码3】: ps.setDouble(3, goodsPrice);
【代码4】: ps.setString(4, goodsType);

6.8.4 实践环节

编写两个 JSP 页面:inputPrepareQuery.jsp 和 showPrepareBy.jsp。用户可以在页面 inputPrepareQuery.jsp 中输入查询条件后,单击"查询"按钮。然后,在 showPrepareBy.jsp 页面中显示符合查询条件的商品信息。在本节任务的 bean(UsePrepare.java)中添加一个方法 getQueryPrepareResultBy() 实现该题的条件查询功能(使用预处理语句实现查询)。页面运行效果如图 6.23 所示。

(a) 使用预处理条件查询

(b) 符合查询条件的记录

图 6.23 页面运行效果

6.9 小 结

本章重点介绍了 JDBC 连接数据库的两种常用方式:建立 JDBC-ODBC 桥接器和加载纯 Java 驱动程序。使用加载纯 Java 驱动程序的方式连接 MySQL、SQL Server 以及 Oracle 等主流关系数据库的步骤基本一样,这些主流关系数据库的新增、更新、删除与查询等操作也基本一样。

习 题 6

1. 当要在 JSP 文件中编写代码连接数据库时,应在 JSP 文件中加入以下(   )语句。
   A. <jsp:include file="java.util.*"/>
   B. <%@ page import="java.sql.*" %>
   C. <jsp:include page="java.lang.*"/>
   D. <%@ page import="java.util.*" %>

2. Java 程序连接数据库常用的两种方式:建立 JDBC-ODBC 和加载纯(   )驱动程序。
   A. Oracle   B. Java   C. Java 数据库   D. 以上都不对

3. 从"员工"表的"姓名"字段找出名字包含"玛丽"的人,下面 select 语句正确的是(   )。
   A. select * from 员工 where 姓名 like '%玛丽%'
   B. select * from 员工 where 姓名 = '%玛丽_'
   C. select * from 员工 where 姓名 like '_玛丽%'
   D. select * from 员工 where 姓名 = '_玛丽_'

4. 下述选项中不属于 JDBC 基本功能的是(   )。
   A. 与数据库建立连接   B. 提交 SQL 语句
   C. 处理查询结果   D. 数据库维护管理

5. 请选出微软公司提供的连接 SQL Server 2005 的 JDBC 驱动程序(   )。
   A. com.mysql.jdbc.Driver
   B. sun.jdbc.odbc.JdbcOdbcDriver
   C. oracle.jdbc.driver.OracleDriver
   D. com.microsoft.sqlserver.jdbc.SQLServerDriver

6. 下面(   )不是 ResultSet 接口的方法。
   A. next()   B. getString()   C. back()   D. getInt()

7. JDBC 能完成哪些工作?

8. 使用纯 Java 数据库驱动程序访问数据库时,都有哪些步骤?

9. JDBC 连接数据库常用的方式有哪些?

第 7 章 Java Servlet

主要内容

(1) servlet 对象的创建与运行。
(2) Servlet 的生命周期。
(3) 通过 JSP 页面访问 Servlet。
(4) doGet()和 doPost()方法。
(5) 重定向与转发。
(6) 会话管理。
(7) 基于 Servlet 的 MVC 模式。

Java Servlet 的核心思想就是在 Web 服务器端创建用来响应用户请求的对象,该对象被称为一个 servlet 对象。JSP 技术以 Java Servlet 为基础,当客户请求一个 JSP 页面时,Web 服务器(如 Tomcat 服务器)会自动生成一个对应的 Java 文件,编译该 Java 文件,并用编译得到的字节码文件在服务器端创建一个 servlet 对象。但实际的 Web 应用需要 servlet 对象具有特定的功能,这时就需要 Web 开发人员编写创建 servlet 对象的类。应如何编写 Servlet 类,又应如何使用 Servlet 类,将在本章中重点介绍。

本章涉及的 Java 源文件保存在工程 ch7 的 src 目录中,涉及的 JSP 页面保存在工程 ch7 的 WebContent 目录中。

7.1 Servlet 类与 servlet 对象

7.1.1 核心知识

编写一个 Servlet 类很简单,只要继承 javax.servlet.http 包中的 HttpServlet 类,并重写响应 HTTP 请求的方法即可。HttpServlet 类实现了 Servlet 接口,实现了响应用户请求的接口方法。对于 HttpServlet 类的一个子类习惯上称为一个 Servlet 类,使用这样的子类创建的对象又习惯上称为 servlet 对象。

7.1.2 能力目标

理解 Servlet 类与 servlet 对象的概念。

7.1.3 任务驱动

1. 任务的主要内容

编写一个简单的 Servlet 类 FirstServlet,用户请求这个 servlet 对象时,就会在浏览器中看到"第一个 Servlet 类"这样的响应信息。

2. 任务的代码模板

FirstServlet.java

```java
package servlet;
import java.io.*;
import javax.servlet.*;
import javax.servlet.http.*;
public class FirstServlet extends 【代码1】{
    private static final long serialVersionUID = 1L;
    public void init(ServletConfig config) throws ServletException{
        super.init(config);
    }
    public void service(HttpServletRequest request,HttpServletResponse response)
            throws IOException{
        //设置响应的内容类型
        response.setContentType("text/html;charset=utf-8");
        //取得输出对象
        PrintWriter out = response.getWriter();
        out.println("<html><body>");
        //在浏览器中显示:第一个 Servlet 类
        out.println("第一个 Servlet 类");
        out.println("</body></html>");
    }
}
```

3. 任务小结或知识扩展

编写 Servlet 类时,必须有包名。也就是说,必须在包中编写 Servlet 类。在本章新建一个 Web 工程 ch7,所有的 Servlet 类放在 src 目录下 servlet 包中。

本任务的 Servlet 类的源文件:FirstServlet.java 保存在 Eclipse 的 Web 工程 ch7 的 src 目录下的 servlet 包中。FirstServlet.java 源文件由 Eclipse 自动编译生成字节码文件 FirstServlet.class,保存在 build\classes\servlet 里面。Servlet 类的源文件与字节码文件保存目录如图 7.1 所示。

编写完 Servlet 类的源文件,并编译了源文件,是否就可以运行 servlet 对象呢? 还不可以,需要部署 Servlet 后,才可以运行 servlet 对象。

4. 代码模板的参考答案

【代码1】:HttpServlet

7.1.4 实践环节

编写一个简单的 Servlet 类 YourFirstServlet。用户请求这个 servlet 对象时,就会在浏

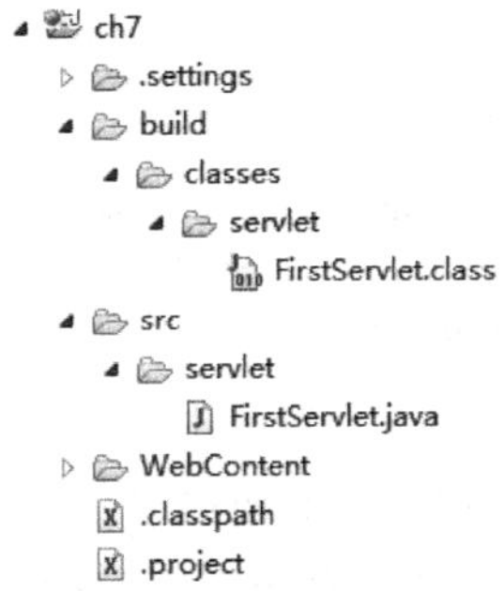

```
▲ 🗁 ch7
    ▷ 🗁 .settings
    ▲ 🗁 build
        ▲ 🗁 classes
            ▲ 🗁 servlet
                🗎 FirstServlet.class
    ▲ 🗁 src
        ▲ 🗁 servlet
            🗎 FirstServlet.java
    ▷ 🗁 WebContent
    🗎 .classpath
    🗎 .project
```

图 7.1　Servlet 类的源文件与字节码文件保存目录

览器中看到"您人生中第一个 Servlet 类"这样的响应信息,并在 Web 工程中找到该 Servlet 类对应的字节码文件。

7.2　servlet 对象的创建与运行

7.2.1　核心知识

要想让 Web 服务器使用 Servlet 类编译后的字节码文件创建 servlet 对象处理用户请求,必须先为 Web 服务器部署 Servlet。部署 Servlet,目前有两种方式:一是在 web.xml 中部署 Servlet;二是基于注解的方式部署 Servlet。

1. 在 web.xml 中部署 Servlet

web.xml 文件由 Web 服务器负责管理,该文件是 Web 应用的部署描述文件,包含如何将用户请求 URL 映射到 Servlet。在 Web 工程的 WebContent\WEB-INF 目录下找到 web.xml 文件,并部署自己的 Servlet。

1) 部署 Servlet

为了在 web.xml 文件里部署 7.1 节中的 FirstServlet,需要在 web.xml 文件里找到 <web-app></web-app>标记,然后在<web-app></web-app>标记中添加如下内容:

```
<servlet>
    <servlet-name>firstServlet</servlet-name>
    <servlet-class>servlet.FirstServlet</servlet-class>
</servlet>
<servlet-mapping>
    <servlet-name>firstServlet</servlet-name>
    <url-pattern>/firstServlet</url-pattern>
</servlet-mapping>
```

2) 运行 Servlet

Servlet 第一次被访问时,需要把它发布到 Web 服务器(选中 Servlet 类的源文件,右击,在弹出的快捷菜单中,选择 Run As/Run on Server 命令)。这时,在 Eclipse 内嵌的浏览器中可看到如图 7.2 所示的运行效果。

把 Servlet 发布到 Web 服务器后,也可以在浏览器的地址栏中输入:

```
← → ■  http://localhost:8080/ch7/firstServlet
第一个Servlet类
```

图 7.2　第一个 Servlet 的运行效果

http://localhost:8080/ch7/firstServlet

来运行 Servlet。

3) web.xml 文件中与 Servlet 部署有关的标记及其说明

① 根标记 web-app。XML 文件中必须有一个根标记，web.xml 的根标记是 web-app。

② servlet 标记及其子标记。web.xml 文件中可以有若干个 servlet 标记，该标记的内容由 Web 服务器负责处理。servlet 标记中有两个子标记：servlet-name 和 servlet-class，其中 servlet-name 子标记的内容是 Web 服务器创建的 servlet 对象的名字。web.xml 文件中可以有若干个 servlet 标记，但要求它们的 servlet-name 子标记的内容互不相同。servlet-class 子标记的内容指定 Web 服务器用哪个类来创建 servlet 对象，如果 servlet 对象已经创建，那么 Web 服务器就不再使用指定的类创建。

③ servlet-mapping 标记及其子标记。web.xml 文件中出现一个 servlet 标记就会对应地出现一个 servlet-mapping 标记。servlet-mapping 标记中也有两个子标记：servlet-name 和 url-pattern。其中 servlet-name 子标记的内容是 Web 服务器创建的 servlet 对象的名字（该名字必须和 servlet 标记的子标记 servlet-name 的内容相同）；url-pattern 子标记用来指定用户用怎样的模式请求 servlet 对象，比如，url-pattern 子标记的内容是/firstServlet，用户要请求服务器运行 servlet 对象 firstServlet 为其服务，那么可在浏览器的地址栏中输入：

http://localhost:8080/ch7/firstServlet

一个 Web 工程的 web.xml 文件负责管理该 Web 工程的所有 servlet 对象，当 Web 工程需要提供更多 servlet 对象时，只要在 web.xml 文件中添加 servlet 和 servlet-mapping 标记即可。

2. 基于注解的方式部署 Servlet

从 7.2.1 节可知，每开发一个 Servlet，需要在 web.xml 中部署 Servlet 才能够使用。这样会给 Web 工程的维护带来非常大的麻烦。在 Servlet 3.0 中提供了注解@WebServlet，不再需要在 web.xml 文件中进行 Servlet 的部署描述，简化开发流程。本书后续 Servlet 都是基于注解的方式部署。

注解虽然方便了开发人员，但在后期会让维护和调试成本增加。为了方便后期维护，建议开发人员部署 Servlet 时把@WebServlet 的属性 urlPatterns 的值设置为 Servlet 类的名字。例如：

```
@WebServlet(name = "secondServlet", urlPatterns = { "/secondServlet" })
public class SecondServlet extends HttpServlet {
}
```

1) @WebServlet 注解

@WebServlet 用于将一个类声明为 Servlet，该注解将会在部署时被 Web 容器处理，Web 容器根据具体的属性将相应的类部署为 Servlet。该注解具有一些常用属性，如表 7.1 所示。

表 7.1 @WebServlet 的常用属性

属性名	类 型	描 述
name	String	指定 Servlet 的 name 属性，等价于＜servlet-name＞，如果没有显式指定，则该 Servlet 的取值即为类的全名
value	String[]	该属性等价于 urlPatterns 属性，两个属性不能同时使用
urlPatterns	String[]	指定一组 Servlet 的 URL 匹配模式，等价于 url-pattern 标记
loadOnStartup	int	指定 Servlet 的加载顺序，等价于 load-on-startup 标记
initParams	WebInitParam[]	指定一组 Servlet 初始化参数，等价于 init-param 标记

以上所有属性均为可选属性，但是 value 或者 urlPatterns 通常是必需的，且两者不能共存，如果同时指定，通常是忽略 value 的取值。

【例 7.1】 基于注解的 Servlet 类——SecondServlet。

SecondServlet.java 的代码如下：

```
package servlet;
import java.io.IOException;
import java.io.PrintWriter;
import javax.servlet.ServletConfig;
import javax.servlet.ServletException;
import javax.servlet.annotation.WebServlet;
import javax.servlet.http.HttpServlet;
import javax.servlet.http.HttpServletRequest;
import javax.servlet.http.HttpServletResponse;
//建议 urlPatterns 的值和类名一样，方便维护
@WebServlet(name = "secondServlet", urlPatterns = { "/secondServlet" })
public class SecondServlet extends HttpServlet {
    private static final long serialVersionUID = 1L;
    public void init(ServletConfig config) throws ServletException {
    }
    protected void service(HttpServletRequest request, HttpServletResponse response) throws
    ServletException, IOException {
        //设置响应的内容类型
        response.setContentType("text/html;charset = utf - 8");
        //取得输出对象
        PrintWriter out = response.getWriter();
        out.println("< html >< body >");
        //在浏览器中显示：第二个 Servlet 类
        out.println("第二个 Servlet 类");
        out.println("</body ></html >");
    }
}
```

SecondServlet.java 代码中使用@WebServlet(name＝"secondServlet"，urlPatterns＝

{ "/secondServlet" })如此部署之后,就不必在 web.xml 中部署相应的 servlet 和 servlet-mapping 元素了,Web 容器会在部署时根据指定的属性将该类发布为 Servlet。@WebServlet(name = "secondServlet", urlPatterns = { "/secondServlet" })等价的 web.xml 部署形式如下:

```
<servlet>
    <servlet-name>secondServlet</servlet-name>
    <servlet-class>servlet.SecondServlet</servlet-class>
</servlet>
<servlet-mapping>
    <servlet-name>secondServlet</servlet-name>
    <url-pattern>/secondServlet</url-pattern>
</servlet-mapping>
```

2) @WebInitParam 注解

@WebInitParam 注解通常不单独使用,而是配合@WebServlet 和@WebFilter(第 8 章讲解)使用。它的作用是为 Servlet 或 Filter 指定初始化参数,这等价于 web.xml 中 servlet 的init-param 子标记。@WebInitParam 具有一些常用属性,如表 7.2 所示。

表 7.2 @WebInitParam 常用属性

属性名	类型	是否可选	描述
name	String	否	指定参数的名字,等价于 param-name
value	String	否	指定参数的值,等价于 param-value

@WebInitParam 注解的示例代码如下:

@WebServlet(name = "thirdServlet", urlPatterns = { "/thirdServlet" }, initParams = {@WebInitParam(name = "firstParam", value = "one"),@WebInitParam(name = "secondParam", value = "two")})

7.2.2 能力目标

掌握部署 Servlet 的两种方法。

7.2.3 任务驱动

1. 任务的主要内容

首先,将 web.xml 文件中有关 Servlet 的部署代码删除,然后使用注解的方式部署 7.1 节的 FirstServlet,并运行部署后的 Servlet。

2. 任务的代码模板

带注解的 FirstServlet.java 的代码模板如下:

```
package servlet;
import java.io.*;
import javax.servlet.*;
import javax.servlet.annotation.WebServlet;
```

```
import javax.servlet.http.*;
【代码1】(name = "firstServlet",【代码2】= { "/firstServlet" })
public class FirstServlet extends HttpServlet {
    private static final long serialVersionUID = 1L;
    public void init(ServletConfig config) throws ServletException{
        super.init(config);
    }
    public void service(HttpServletRequest request,HttpServletResponse response)
    throws IOException{
        //设置响应的内容类型
        response.setContentType("text/html;charset = utf - 8");
        //取得输出对象
        PrintWriter out = response.getWriter();
        out.println("<html><body>");
        //在浏览器中显示：人生第一个 Servlet 类
        out.println("人生第一个 Servlet 类");
        out.println("</body></html>");
    }
}
```

3. 任务小结或知识扩展

一个 servlet 对象的生命周期主要由下列 3 个过程组成。

1) 初始化 servlet 对象

当 servlet 对象第一次被请求加载时，服务器会创建一个 servlet 对象，该 servlet 对象调用 init()方法完成必要的初始化工作。

2) service()方法响应请求

创建的 servlet 对象再调用 service()方法响应客户的请求。

3) servlet 对象死亡

当服务器关闭时，servlet 对象调用 destroy()方法使自己消亡。

从上面 3 个过程来看，init()方法只能被调用一次，即在 Servlet 第一次被请求加载时调用该方法。当客户请求 Servlet 服务时，服务器将启动一个新的线程，在该线程中，servlet 对象调用 service()方法响应客户的请求。那么多个客户请求 Servlet 服务时，服务器会怎么办呢？服务器会为每个客户启动一个新的线程，在每个线程中，servlet 对象调用 service()方法响应客户的请求。也就是说，每个客户请求都会导致 service()方法被调用执行，分别运行在不同的线程中。

【例 7.2】 Servlet 接口的 init()、service()和 destroy()方法。

ThirdServlet.java 的代码如下：

```
package servlet;
import java.io.IOException;
import javax.servlet.ServletConfig;
import javax.servlet.ServletException;
import javax.servlet.annotation.WebInitParam;
import javax.servlet.annotation.WebServlet;
import javax.servlet.http.HttpServlet;
import javax.servlet.http.HttpServletRequest;
```

```java
import javax.servlet.http.HttpServletResponse;
@WebServlet(name = "thirdServlet", urlPatterns = { "/thirdServlet" },
initParams = {@WebInitParam(name = "firstParam", value = "one"),
        @WebInitParam(name = "secondParam", value = "two")})
public class ThirdServlet extends HttpServlet {
    private static final long serialVersionUID = 1L;
    private String first = null;
    private String second = null;
    private static int count = 0;
    public void init(ServletConfig config) throws ServletException {
        //获得参数 firstParam 的值
        first = config.getInitParameter("firstParam");
        second = config.getInitParameter("secondParam");
        System.out.println("第一个参数值:" + first);
        System.out.println("第二个参数值:" + second);
    }
    protected void service(HttpServletRequest request, HttpServletResponse response) throws ServletException, IOException {
        count ++;
        System.out.println("您是第" + count + "个客户请求该Servlet!");
    }
    public void destroy() {
    }
}
```

在 ThirdServlet 的 init()方法中,通过 ServletConfig 的对象 config 调用 getInitParameter() 方法来获取参数的值。请求该 Servlet 3 次后,在 Eclipse 控制台中打印出如图 7.3 所示结果。

```
❀ Servers  🗏 Console ✕
第一个参数值:one
第二个参数值:two
您是第1个客户请求该Servlet!
您是第2个客户请求该Servlet!
您是第3个客户请求该Servlet!
```

图 7.3　请求 3 次 thirdServlet 的结果

从图 7.3 看出,不管请求几次 thirdServlet,它的 init()方法只执行一次。而 service()方法每请求一次,就执行一次。

4. 代码模板的参考答案

【代码 1】: @WebServlet
【代码 2】: urlPatterns

7.2.4　实践环节

编写一个简单的 Servlet 类 Pratice2Servlet,使用注解的方式部署该 servlet 对象并运行它。用户通过在 IE 浏览器地址栏中输入 http://localhost:8080/ch7/pratice2 请求这个 servlet 对象时,就会在浏览器中看到"部署与运行 servlet"的响应信息。

7.3 通过 JSP 页面访问 Servlet

7.3.1 核心知识

可以通过 JSP 页面的表单或超链接请求某个 Servlet。通过 JSP 页面访问 Servlet 的好处是,JSP 页面负责页面的静态信息处理,动态信息处理由 Servlet 完成。

1. 通过表单访问 Servlet

假设在 JSP 页面中,有如下表单:

```
<form action = "isLogin" method = "post">
    …
</form>
```

那么该表单的处理程序(action)就是一个 Servlet,为该 Servlet 部署时,@WebServlet 的 urlPatterns 属性值为{ "/isLogin" }。

2. 通过超链接访问 Servlet

在 JSP 页面中,可以单击超链接,访问 servlet 对象,同时也可以通过超链接向 Servlet 提交信息,例如,<a href="loginServlet? user=taipingle&&pwd=zhenzuile">查看用户名和密码</a>,"查看用户名和密码"这个超链接就将 user=taipingle 和 pwd=zhenzuile 两个信息提交给 Servlet 处理。

7.3.2 能力目标

能够灵活使用 JSP 页面访问 servlet 对象。

7.3.3 任务驱动

1. 任务的主要内容

编写一个 JSP 页面 login.jsp,在该页面中通过表单向 urlPatterns 为{ "/loginServlet" }的 Servlet(由 LoginServlet 类负责创建)提交用户名和密码,Servlet 负责判断输入的用户名(zhangsan)和密码(lisi)是否正确,并将判断结果返回。运行效果如图 7.4 所示。

2. 任务的代码模板

页面文件 login.jsp 的代码如下:

```
<%@ page language="java" contentType="text/html; charset=GBK" pageEncoding="GBK"%>
<html>
    <head>
        <title>login.jsp</title>
    </head>
    <body>
        <form action="loginServlet" method="post">
            <table>
                <tr>
                    <td>用户名:</td>
```

(a) 信息输入页面

(b) 信息正确页面

(c) 信息错误页面

图 7.4　页面运行效果

```
            <td><input type = "text" name = "user"/></td>
        </tr>
        <tr>
            <td>密 码：</td>
            <td><input type = "password" name = "pwd"/></td>
        </tr>
        <tr>
            <td><input type = "submit" value = "提交"/></td>
            <td><input type = "reset" value = "重置"/></td>
        </tr>
    </table>
    </form>
  </body>
</html>
```

LoginServlet.java 的代码如下：

```
package servlet;
import java.io.IOException;
import java.io.PrintWriter;
import javax.servlet.ServletException;
import javax.servlet.annotation.WebServlet;
import javax.servlet.http.HttpServlet;
import javax.servlet.http.HttpServletRequest;
import javax.servlet.http.HttpServletResponse;
```
【代码 1】//使用注解的方式部署该 Servlet
```
public class LoginServlet extends HttpServlet {
    private static final long serialVersionUID = 1L;
    protected void service(HttpServletRequest request, HttpServletResponse response) throws
    ServletException, IOException {
        response.setContentType("text/html;charset = GBK");
        PrintWriter out = response.getWriter();
        request.setCharacterEncoding("GBK");            //设置编码,防止中文乱码
        String name = request.getParameter("user");     //获取客户提交的信息
        String password = request.getParameter("pwd");  //获取客户提交的信息
```

```
            out.println("<html><body>");
            if(name == null || name.length() == 0){
                out.println("请输入用户名");
            }
            else if(password == null || password.length() == 0){
                out.println("请输入密码");
            }
            else if(name.length() > 0 && password.length() > 0){
                if(name.equals("zhangsan") && password.equals("lisi")){
                    out.println("信息输入正确");
                }else{
                    out.println("信息输入错误");
                }
            }
            out.println("</body></html>");
        }
    }
```

3. 任务小结或知识扩展

需要特别注意的是，如果 Servlet 的请求格式是/×××（请求格式就是 urlPatterns 的值），那么 JSP 页面请求 Servlet 时，必须写成×××，不可以写成/×××，否则将变成请求服务器(Tomcat)root 目录下的某个 Servlet。

4. 代码模板的参考答案

【代码 1】：`@WebServlet(name = "loginServlet", urlPatterns = { "/loginServlet" })`

7.3.4 实践环节

将 7.3.3 节任务中通过表单访问 LoginServlet 改成通过超链接方式访问。

7.4 doGet 和 doPost 方法

7.4.1 核心知识

当服务器接收到一个 Servlet 请求时，就会产生一个新线程，在这个线程中让 servlet 对象调用 service()方法为请求做出响应。service()方法首先检查 HTTP 请求类型(get 或 post)，并在 service()方法中根据用户的请求方式，对应地再调用 doGet()方法或 doPost()方法。

HTTP 请求类型为 get 方式时，service()方法调用 doGet()方法响应用户请求；HTTP 请求类型为 post 方式时，service()方法调用 doPost()方法响应用户请求。因此，在 Servlet 类中，没有必要重写 service()方法，直接继承即可。

在 Servlet 类中重写 doGet()或 doPost()方法来响应用户的请求，这样可以增加响应的灵活性，同时减轻服务器的负担。

7.4.2 能力目标

理解 doGet()方法和 doPost()方法的调用原理。

7.4.3 任务驱动

1. 任务的主要内容

编写一个 JSP 页面 inputLader.jsp，在该页面中使用表单向 urlPatterns 为{ "/ getLengthOrAreaServlet" }的 Servlet 提交矩形的长与宽。Servlet（由 GetLengthOrAreaServlet 负责创建）处理手段依赖表单提交数据的方式，当提交方式为 get 时，Servlet 计算矩形的周长，当提交方式为 post 时，Servlet 计算矩形的面积。页面运行效果如图 7.5 所示。

（a）信息输入页面

（b）post 方式提交获得矩形面积

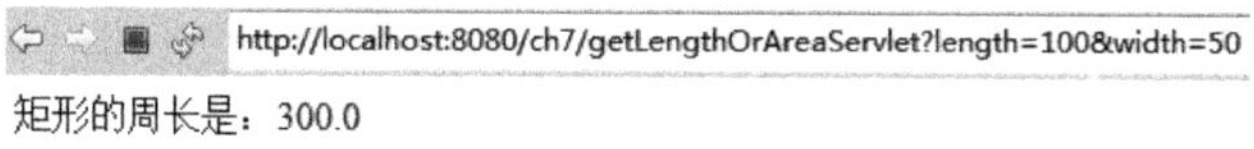

（c）get 方式提交获得矩形周长

图 7.5　页面运行效果

2. 任务的代码模板

页面文件 inputLader.jsp 的代码如下：

```
<%@ page language="java" contentType="text/html; charset=GBK" pageEncoding="GBK"%>
<html>
    <head>
        <title>inputLader.jsp</title>
    </head>
    <body>
        <h2>输入矩形的长和宽,提交给 servlet(post 方式)求面积:</h2>
        【代码1】
        长:<input type="text" name="length"/><br/>
        宽:<input type="text" name="width"/><br/>
            <input type="submit" value="提交"/>
        </form>
            <br/>
        <h2>输入矩形的长和宽,提交给 servlet(get 方式)求周长:</h2>
            【代码2】
```

```html
            长: <input type = "text" name = "length"/><br/>
            宽: <input type = "text" name = "width"/><br/>
            <input type = "submit" value = "提交"/>
        </form>
    </body>
</html>
```

GetLengthOrAreaServlet.java 的代码如下:

```java
package servlet;
import java.io.IOException;
import java.io.PrintWriter;
import javax.servlet.ServletException;
import javax.servlet.annotation.WebServlet;
import javax.servlet.http.HttpServlet;
import javax.servlet.http.HttpServletRequest;
import javax.servlet.http.HttpServletResponse;
@WebServlet(name = "getLengthOrAreaServlet", urlPatterns = { "/getLengthOrAreaServlet" })
public class GetLengthOrAreaServlet extends HttpServlet {
    private static final long serialVersionUID = 1L;
    protected void doGet(HttpServletRequest request, HttpServletResponse response) throws ServletException, IOException {
        response.setContentType("text/html;charset=utf-8");
        PrintWriter out = response.getWriter();
        String l = request.getParameter("length");
        String w = request.getParameter("width");
        out.println("<html><body>");
        double m = 0 ,n = 0;
        try{
            m = Double.parseDouble(l);
            n = Double.parseDouble(w);
            out.println("矩形的周长是: " + ( m + n ) * 2);
        }catch(NumberFormatException e){
            out.println("请输入数字字符!");
        }
        out.println("</body></html>");
    }
    protected void doPost(HttpServletRequest request, HttpServletResponse response) throws ServletException, IOException {
        response.setContentType("text/html;charset=GBK");
        PrintWriter out = response.getWriter();
        String l = request.getParameter("length");
        String w = request.getParameter("width");
        out.println("<html><body>");
        double m = 0, n = 0;
        try{
            m = Double.parseDouble(l);
            n = Double.parseDouble(w);
            out.println("矩形的面积是: " + m * n);
        }catch(NumberFormatException e){
            out.println("请输入数字字符!");
```

```
            }
            out.println("</body></html>");
        }
    }
```

3. 任务小结或知识扩展

一般情况下,如果不论用户请求类型是 get 还是 post,服务器的处理过程完全相同,那么可以只在 doPost()方法中编写处理过程,而在 doGet()方法中再调用 doPost()方法;或只在 doGet()方法中编写处理过程,而在 doPost()方法中再调用 doGet()方法。

4. 代码模板的参考答案

【代码 1】: < form action = "getLengthOrAreaServlet" method = "post">
【代码 2】: < form action = "getLengthOrAreaServlet" method = "get">

7.4.4 实践环节

编写一个 JSP 页面 pratice7_4.jsp,在该页面中使用表单向 urlPatterns 为{"/praticeServlet"}的 Servlet 提交矩形的长与宽。Servlet(由 PraticeServlet 类负责创建)处理手段不依赖于表单提交方式,即不论 post 还是 get,处理数据的手段相同,都是计算矩形的周长。

7.5 重定向与转发

重定向是将用户从当前 JSP 页面或 Servlet 定向到另一个 JSP 页面或 Servlet,以前的 request 中存放的信息全部失效,并进入一个新的 request 作用域;转发是将用户对当前 JSP 页面或 Servlet 的请求转发给另一个 JSP 页面或 Servlet,以前的 request 中存放的信息不会失效。

7.5.1 核心知识

1. 重定向

在 Servlet 中通过调用 HttpServletResponse 类中的方法 sendRedirect(String location)来实现重定向,重定向的目标页面或 Servlet(由参数 location 指定),无法从以前的 request 对象中获取用户提交的数据。

2. 转发

javax.servlet.RequestDispatcher 对象可以将用户对当前 JSP 页面或 Servlet 的请求转发给另一个 JSP 页面或 Servlet。实现转发需要以下两个步骤。

1) 获得 RequestDispatcher 对象

在当前 JSP 页面或 Servlet 中,使用 request 对象调用

```
public RequestDispatcher getRequestDispatcher(String url)
```

方法返回一个 RequestDispatcher 对象,其中参数 url 就是要转发的 JSP 页面或 Servlet 的

地址，例如：

```
RequestDispatcher dis = request.getRequestDispatcher("dologin");
```

2）RequestDispatcher 对象调用 forward()方法实现转发

获得 RequestDispatcher 对象之后，就可以使用该对象调用

```
public void forward(ServletRequest request, ServletResponse response)
```

方法将用户对当前 JSP 页面或 Servlet 的请求转发给 RequestDispatcher 对象所指定的 JSP 页面或 Servlet。例如：

```
dis.forward(request, response);
```

7.5.2 能力目标

理解重定向与转发的区别，掌握重定向与转发的实现方法。

7.5.3 任务驱动

1. 任务的主要内容

编写 JSP 页面 redirectForward.jsp，在该 JSP 页面中通过表单向 urlPatterns 为 {"/redirectForwardServlet"}的 Servlet（由 RedirectForwardServlet 负责创建）提交用户名和密码。如果用户输入的数据不完整，redirectForwardServlet 将用户重定向到 redirectForward.jsp 页面；如果用户输入的数据完整，redirectForwardServlet 将用户对 redirectForward.jsp 页面的请求转发给 urlPatterns 为{"/showServlet"}的 Servlet（由 ShowServlet 负责创建），showServlet 显示用户输入的信息。

2. 任务的代码模板

页面文件 redirectForward.jsp 的代码如下：

```jsp
<%@ page language="java" contentType="text/html; charset=GBK" pageEncoding="GBK"%>
<html>
    <head>
        <title>redirectForward.jsp</title>
    </head>
    <body>
        <form action="redirectForwardServlet" method="post">
            <table>
                <tr>
                    <td>用户名：</td>
                    <td><input type="text" name="user"/></td>
                </tr>
                <tr>
                    <td>密 码：</td>
                    <td><input type="password" name="pwd"/></td>
                </tr>
                <tr>
                    <td><input type="submit" value="提交"/></td>
```

```html
            <td><input type="reset" value="重置"/></td>
          </tr>
        </table>
      </form>
    </body>
</html>
```

RedirectForwardServlet.java 的代码如下:

```java
package servlet;
import java.io.IOException;
import javax.servlet.RequestDispatcher;
import javax.servlet.ServletException;
import javax.servlet.annotation.WebServlet;
import javax.servlet.http.HttpServlet;
import javax.servlet.http.HttpServletRequest;
import javax.servlet.http.HttpServletResponse;
@WebServlet(name = "redirectForwardServlet", urlPatterns = { "/redirectForwardServlet" })
public class RedirectForwardServlet extends HttpServlet {
    private static final long serialVersionUID = 1L;
    protected void doGet(HttpServletRequest request,
            HttpServletResponse response) throws ServletException, IOException {
        doPost(request, response);
    }
    protected void doPost(HttpServletRequest request,
            HttpServletResponse response) throws ServletException, IOException {
        request.setCharacterEncoding("GBK");
        String name = request.getParameter("user");
        String password = request.getParameter("pwd");
        if (name == null || name.length() == 0) {
            //使用 response 调用 sendRedirect 方法重定向到 redirectForward.jsp
            【代码 1】
        } else if (password == null || password.length() == 0) {
            【代码 2】
        } else if (name.length() > 0 && password.length() > 0) {
            //转发
            【代码 3】
            【代码 4】
        }
    }
}
```

ShowServlet.java 的代码如下:

```java
package servlet;
import java.io.IOException;
import java.io.PrintWriter;
import javax.servlet.ServletException;
import javax.servlet.annotation.WebServlet;
import javax.servlet.http.HttpServlet;
import javax.servlet.http.HttpServletRequest;
import javax.servlet.http.HttpServletResponse;
```

```java
@WebServlet(name = "showServlet", urlPatterns = { "/showServlet" })
public class ShowServlet extends HttpServlet {
    private static final long serialVersionUID = 1L;
    protected void doGet(HttpServletRequest request,
            HttpServletResponse response) throws ServletException, IOException {
        doPost(request, response);
    }
    protected void doPost(HttpServletRequest request,
            HttpServletResponse response) throws ServletException, IOException {
        response.setContentType("text/html;charset = GBK");
        PrintWriter out = response.getWriter();
        request.setCharacterEncoding("GBK");
        String name = request.getParameter("user");
        String password = request.getParameter("pwd");
        out.println("您的用户名是: " + name);
        out.println("<br>您的密码是: " + password);
    }
}
```

3. 任务小结或知识扩展

转发是服务器行为,重定向是客户端行为。具体工作流程如下。

转发过程:客户浏览器发送 HTTP 请求,Web 服务器接收此请求,调用内部的一个方法在容器内部完成请求处理和转发动作,将目标资源发送给客户;在这里,转发的路径必须是同一个 Web 容器下的 URL,其不能转向到其他的 Web 路径中,中间传递的是自己的容器内的 request。在客户浏览器的地址栏中显示的仍然是其第一次访问的路径,也就是说客户是感觉不到服务器做了转发的。转发行为是浏览器只做了一次访问请求。

重定向过程:客户浏览器发送 HTTP 请求,Web 服务器接收后发送 302 状态码响应及对应新的 location 给客户浏览器,客户浏览器发现是 302 响应,则自动再发送一个新的 HTTP 请求,请求 URL 是新的 location 地址,服务器根据此请求寻找资源并发送给客户。在这里 location 可以重定向到任意 URL,既然是浏览器重新发出了请求,就没有什么 request 传递的概念了。在客户浏览器的地址栏中显示的是其重定向的路径,客户可以观察到地址的变化。重定向行为是浏览器做了至少两次的访问请求。

4. 代码模板的参考答案

【代码 1】: response.sendRedirect("redirectForward.jsp");
【代码 2】: response.sendRedirect("redirectForward.jsp");
【代码 3】: RequestDispatcher dis = request.getRequestDispatcher("showServlet");
【代码 4】: dis.forward(request, response);

7.5.4 实践环节

试将任务中的转发(代码 3 与代码 4 部分)改成重定向,然后运行 redirectForward.jsp 页面,看看运行结果是什么样的,为什么是这样的结果?

7.6 在 Servlet 中使用 session

7.6.1 核心知识

在 Servlet 中获得当前请求的会话对象可通过调用 HttpServletRequest 的 getSession()方法实现,例如：

```
HttpSession session = request.getSession(true);  //若存在会话则返回该会话,否则新建一个
                                                  //会话
```

或

```
HttpSession session = request.getSession(false); //若存在会话则返回该会话,否则返回null
```

经常情况下,通过第一种方式获得 Session,即指定 getSession()的参数为 true。默认参数为 true,即 request.getSession(true)等同于 request.getSession()。

7.6.2 能力目标

掌握如何在 Servlet 中使用会话对象 session。

7.6.3 任务驱动

1. 任务的主要内容

编写一个 JSP 页面 useSession.jsp,在该页面中通过表单向 urlPatterns 为 {"/useSessionServlet"}的 servlet 对象(由 UseSessionServlet 类负责创建)提交用户名, useSessionServlet 将用户名存入 session 对象中,然后用户请求另一个 urlPatterns 为 {"/showNameServlet"}的 Servlet(由 ShowNameServlet 类负责创建),showNameServlet 从用户的 session 对象中取出存储的用户名,并显示在浏览器中。程序运行效果如图 7.6 所示。

2. 任务的代码模板

页面文件 useSession.jsp 的代码如下：

```
<%@ page language="java" contentType="text/html; charset=GBK" pageEncoding="GBK"%>
<html>
    <head>
        <title>useSession.jsp</title>
    </head>
    <body>
        <form action="useSessionServlet" method="post">
            <table>
                <tr>
                    <td>用户名：</td>
                    <td><input type="text" name="user"/></td>
                </tr>
                <tr>
                    <td><input type="submit" value="提交"/></td>
```

第 7 章 Java Servlet

(a) 信息输入页面

(b) 获取会话并存储数据

(c) 获取会话中的数据并显示

图 7.6 在 Servlet 中使用 session

```
            </tr>
        </table>
    </form>
  </body>
</html>
```

UseSessionServlet.java 的代码模板如下：

```
package servlet;
import java.io.*;
import javax.servlet.*;
import javax.servlet.annotation.WebServlet;
import javax.servlet.http.*;
@WebServlet(name = "useSessionServlet", urlPatterns = { "/useSessionServlet" })
public class UseSessionServlet extends HttpServlet {
    private static final long serialVersionUID = 1L;
    public void init(ServletConfig config) throws ServletException {
        super.init(config);
    }
    public void doPost(HttpServletRequest request, HttpServletResponse response)
            throws ServletException, IOException {
        response.setContentType("text/html;charset = GBK");
        PrintWriter out = response.getWriter();
        request.setCharacterEncoding("GBK");
        String name = request.getParameter("user");
        if (null == name || name.trim().length() == 0) {
            response.sendRedirect("useSession.jsp");
        } else {
            【代码 1】
            session.setAttribute("myName", name);
            out.println("<html><body>");
```

```java
            out.println("您请求的servlet对象是: " + getServletName());
            out.println("<br>您的会话ID是: " + session.getId());
            out.println("<br>请单击请求另一个servlet: ");
            out.println("<br><a href = 'showNameServlet'>请求另一个servlet</a>");
            out.println("</body></html>");
        }
    }
    public void doGet(HttpServletRequest request, HttpServletResponse response)
            throws ServletException, IOException {
        doPost(request, response);
    }
}
```

ShowNameServlet.java 的代码模板如下：

```java
package servlet;
import java.io.*;
import javax.servlet.*;
import javax.servlet.annotation.WebServlet;
import javax.servlet.http.*;
@WebServlet(name = "showNameServlet", urlPatterns = { "/showNameServlet" })
public class ShowNameServlet extends HttpServlet {
    private static final long serialVersionUID = 1L;
    public void init(ServletConfig config) throws ServletException {
        super.init(config);
    }
    public void doPost(HttpServletRequest request, HttpServletResponse response)
            throws ServletException, IOException {
        response.setContentType("text/html;charset = GBK");
        PrintWriter out = response.getWriter();
        【代码2】
        String name = (String) session.getAttribute("myName");
        if (null == name || name.trim().length() == 0) {
            response.sendRedirect("useSession.jsp");
        }else{
            out.println("<html><body>");
            out.println("您请求的servlet对象是: " + getServletName());
            out.println("<br>您的会话ID是: " + session.getId());
            out.println("<br>您的会话中存储的用户名是: " + name);
            out.println("<br><a href = useSession.jsp>重新登录</a>");
            out.println("</body></html>");
        }
    }
    public void doGet(HttpServletRequest request, HttpServletResponse response)
            throws ServletException, IOException {
        doPost(request, response);
    }
}
```

3. 任务小结或知识扩展

用户的会话对象 session 在 JSP 页面中可以不用声明直接使用，而在 Servlet 类中必须先使用 request 对象获得用户的会话对象,然后再使用它。

4. 代码模板的参考答案

【代码 1】: HttpSession session = request.getSession(true);
【代码 2】: HttpSession session = request.getSession(true);

7.6.4 实践环节

请阐述在 JSP 页面中使用会话对象 session 和在 Servlet 中使用会话对象 session 有什么不同,并举例说明。

7.7 基于 Servlet 的 MVC 模式

将大量的 Java 代码写在 JSP 页面中,将 HTML 代码写在 Servlet 中。这样会造成代码编写不易,日后维护也不易。因此,学习 Web 应用程序的设计模式是非常重要的。本节将学习一种非常典型的 Web 应用程序的设计模式——基于 Servlet 的 MVC 模式。

7.7.1 核心知识

1. MVC 的概念

MVC 是 Model、View、Controller 的缩写,分别代表 Web 应用程序中的 3 种职责。
(1) 模型——用于存储数据以及处理用户请求的业务逻辑。
(2) 视图——向控制器提交数据,显示模型中的数据。
(3) 控制器——根据视图提出的请求,判断将请求和数据交给哪个模型处理,处理后的有关结果交给哪个视图更新显示。

2. 基于 Servlet 的 MVC 模式

基于 Servlet 的 MVC 模式的具体实现如下。
(1) 模型:一个或多个 JavaBean 对象,用于存储数据(实体模型,由 JavaBean 类创建)和处理业务逻辑(业务模型,由一般的 Java 类创建)。
(2) 视图:一个或多个 JSP 页面,向控制器提交数据和为模型提供数据显示,JSP 页面主要使用 HTML 标记和 JavaBean 标记来显示数据。
(3) 控制器:一个或多个 servlet 对象,根据视图提交的请求进行控制,即将请求转发给处理业务逻辑的 JavaBean,并将处理结果存放到实体模型 JavaBean 中,输出给视图显示。

基于 Servlet 的 MVC 模式的流程如图 7.7 所示。

7.7.2 能力目标

掌握基于 Servlet 的 MVC 模式。

7.7.3 任务驱动

1. 任务的主要内容

使用 MVC 模式实现简单的用户登录验证程序,其中包括实体模型 User、业务模型

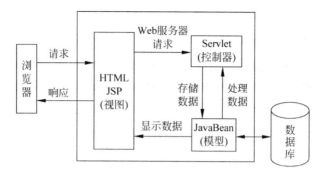

图 7.7 基于 Servlet 的 MVC 模式的流程

UserCheck、控制器 LoginCheckServlet 和两个视图页面,即登录页面和登录成功页面。

2. 任务的代码模板

User 类(实体层)用于创建实体模型存储用户信息,代码如下:

```
package dto;
public class User {
    private String name;
    private String pwd;
    public String getName() {
        return name;
    }
    public void setName(String name) {
        this.name = name;
    }
    public String getPwd() {
        return pwd;
    }
    public void setPwd(String pwd) {
        this.pwd = pwd;
    }
}
```

UserCheck(业务层)类用于判断用户名和密码是否正确,代码如下:

```
package service;
import dto.User;
public class UserCheck {
    //验证登录
    public boolean validate(User user) {
        if (user != null && user.getName().equals("JSPMVC")) {
            if (user.getPwd().equals("MVC")) {
                return true;
            }
            return false;
        }
        return false;
    }
```

}

LoginCheckServlet(控制层)完成请求控制,代码如下:

```java
package servlet;
import java.io.IOException;
import javax.servlet.RequestDispatcher;
import javax.servlet.ServletException;
import javax.servlet.annotation.WebServlet;
import javax.servlet.http.HttpServlet;
import javax.servlet.http.HttpServletRequest;
import javax.servlet.http.HttpServletResponse;
import service.UserCheck;
import dto.User;

@WebServlet(name = "loginCheckServlet", urlPatterns = { "/loginCheckServlet" })
public class LoginCheckServlet extends HttpServlet {
    private static final long serialVersionUID = 1L;
    protected void doGet(HttpServletRequest request,
            HttpServletResponse response) throws ServletException, IOException {
        doPost(request, response);
    }
    protected void doPost(HttpServletRequest request,
            HttpServletResponse response) throws ServletException, IOException {
        request.setCharacterEncoding("GBK");
        String name = request.getParameter("name");
        String pwd = request.getParameter("pwd");
        User user = new User();            //实例化实体模型 User
        user.setName(name);                //把数据存在模型 User 中
        user.setPwd(pwd);                  //把数据存在模型 User 中
        UserCheck uc = new UserCheck();    //实例化业务模型 UserCheck
        if (uc.validate(user)) {
            //把装有数据的实体模型 User,存储到 request 范围内
            request.setAttribute("user", user);
            RequestDispatcher dis = request
                    .getRequestDispatcher("loginSuccess.jsp");
            dis.forward(request, response);
        } else {
            response.sendRedirect("loginCheck.jsp");
        }
    }
}
```

登录页面 loginCheck.jsp()视图层的代码如下:

```jsp
<%@ page language = "java" contentType = "text/html; charset = GBK" pageEncoding = "GBK" %>
<html>
    <head>
        <title>loginCheck.jsp</title>
    </head>
    <body>
        <form action = "loginCheckServlet" method = "post">
```

```
            <table>
                <tr>
                    <td>用户名：</td>
                    <td><input type="text" name="name"/></td>
                </tr>
                <tr>
                    <td>密　码：</td>
                    <td><input type="password" name="pwd"/></td>
                </tr>
                <tr>
                    <td><input type="submit" value="提交"/></td>
                    <td><input type="reset" value="重置"/></td>
                </tr>
            </table>
        </form>
    </body>
</html>
```

登录成功页面 loginSuccess.jsp（视图层）的代码如下：

```
<%@ page language="java" contentType="text/html; charset=GBK" pageEncoding="GBK"%>
<%@ page import="dto.User" %>
<html>
    <head>
        <title>loginSuccess.jsp</title>
    </head>
    <body>
        <jsp:useBean id="user" type="dto.User" scope="request"/>
        恭喜<jsp:getProperty property="name" name="user"/>登录成功！
    </body>
</html>
```

3. 任务小结或知识扩展

在基于 Servlet 的 MVC 模式中，控制器 Servlet 创建的实体模型 JavaBean 也涉及生命周期，生命周期分别为 request、session 和 application。下面以任务中的实体模型 User 来讨论这 3 种生命周期的用法。

1) request 周期的模型

使用 request 周期的模型一般需要以下几个步骤。

① 创建模型并把数据保存到模型中。在 Servlet 中需要以下代码：

```
User user = new User();                    //实例化模型 User
user.setName(name);                        //把数据存在模型 User 中
user.setPwd(pwd);                          //把数据存在模型 User 中
```

② 将模型保存到 request 对象中并转发给视图 JSP。在 Servlet 中需要以下代码：

```
request.setAttribute("user", user);
                        //把装有数据的模型 User 输出给视图 loginSuccess.jsp 页面
RequestDispatcher dis = request.getRequestDispatcher("loginSuccess.jsp");
dis.forward(request, response);
```

request.setAttribute("user"，user)这句代码指定了查找 JavaBean 的关键字，并决定了 JavaBean 的生命周期为 request。

③ 视图更新。Servlet 所转发的页面，比如 loginSuccess.jsp 页面，必须使用 useBean 标记获得 Servlet 所创建的 JavaBean 对象（视图不负责创建 JavaBean）。在 JSP 页面需要使用以下代码：

```
<jsp:useBean id="user" type="dto.User" scope="request"/>
<jsp:getProperty property="name" name="user"/>
```

标记中的 id 就是 Servlet 所创建的模型 JavaBean，它和 request 对象中的关键字对应。因为在视图中不创建 JavaBean 对象，所以在 useBean 标记中使用 type 属性，而不使用 class 属性。useBean 标记中的 scope 必须和存储模型时的范围（request）一致。

2）session 周期的模型

使用 session 周期的模型一般需要以下几个步骤。

① 创建模型并把数据保存到模型中。在 Servlet 中需要以下代码：

```
User user = new User();              //实例化模型 User
user.setName(name);                  //把数据存在模型 User 中
user.setPwd(pwd);                    //把数据存在模型 User 中
```

② 将模型保存到 session 对象中并转发给视图 JSP。在 Servlet 中需要以下代码：

```
session.setAttribute("user", user);  //把装有数据的模型 User 输出给视图
                                     //loginSuccess.jsp 页面
RequestDispatcher dis = request.getRequestDispatcher("loginSuccess.jsp");
dis.forward(request, response);
```

session.setAttribute("user"，user)这句代码指定了查找 JavaBean 的关键字，并决定了 JavaBean 的生命周期为 session。

③ 视图更新。Servlet 所转发的页面，比如 loginSuccess.jsp 页面，必须使用 useBean 标记获得 Servlet 所创建的 JavaBean 对象（视图不负责创建 JavaBean）。在 JSP 页面需要使用以下代码：

```
<jsp:useBean id="user" type="dto.User" scope="session"/>
<jsp:getProperty property="name" name="user"/>
```

标记中的 id 就是 Servlet 所创建的模型 JavaBean，它和 session 对象中的关键字对应。因为在视图中不创建 JavaBean 对象，所以在 useBean 标记中使用 type 属性，而不使用 class 属性。useBean 标记中的 scope 必须和存储模型时的范围（session）一致。

注意：对于生命周期为 session 的模型，Servlet 不仅可以使用 RequestDispatcher 对象转发给 JSP 页面，也可以使用 response 的重定向方法（sendRedirect）定向到 JSP 页面。

3）application 周期的模型

使用 application 周期的模型一般需要以下几个步骤。

① 创建模型并把数据保存到模型中。在 Servlet 中需要以下代码：

```
User user = new User();              //实例化模型 User
user.setName(name);                  //把数据存在模型 User 中
```

```
user.setPwd(pwd);                              //把数据存在模型 User 中
```

② 将模型保存到 application 对象中并转发给视图 JSP。在 Servlet 中需要以下代码：

```
application.setAttribute("user", user);        //把装有数据的模型 User 输出给视图
                                               //loginSuccess.jsp 页面
RequestDispatcher dis = request.getRequestDispatcher("loginSuccess.jsp");
dis.forward(request, response);
```

application.setAttribute("user"，user)这句代码指定了查找 JavaBean 的关键字，并决定了 JavaBean 的生命周期为 application。

③ 视图更新。Servlet 所转发的页面，比如 loginSuccess.jsp 页面，必须使用 useBean 标记获得 Servlet 所创建的 JavaBean 对象(视图不负责创建 JavaBean)。在 JSP 页面需要使用以下代码：

```
<jsp:useBean id = "user" type = "dto.User" scope = "application"/>
<jsp:getProperty property = "name" name = "user"/>
```

标记中的 id 就是 Servlet 所创建的模型 JavaBean，它和 application 对象中的关键字对应。因为在视图中不创建 JavaBean 对象，所以在 useBean 标记中使用 type 属性，而不使用 class 属性。useBean 标记中的 scope 必须和存储模型时的范围(application)一致。

注意：对于生命周期为 session 或 application 的模型，Servlet 不仅可以使用 RequestDispatcher 对象转发给 JSP 页面，也可以使用 response 的重定向方法(sendRedirect())定向到 JSP 页面。

7.7.4 实践环节

模仿 7.7.3 节中的任务，使用基于 Servlet 的 MVC 模式设计一个 Web 应用，要求如下。

用户通过 JSP 页面 inputNumber.jsp 输入两个操作数，并选择一种运算符，单击"提交"按钮后，调用 HandleComputer.java 这个 Servlet。在 HandleComputer 中首先获取用户输入的数字和运算符并将这些内容存入实体模型(由 Computer.java 创建)中，然后调用业务模型(由 CalculateBean.java 创建)进行计算并把结果存入实体模型中，在 showResult.jsp 中调用 JavaBean 显示计算的结果。

7.8 小　　结

本章使用了 Servlet 3.0 的注解机制部署 Servlet，简化了 Servlet 的开发流程，使 web.xml 部署描述文件从 Servlet 3.0 开始不再是必选的。

习　题　7

1. 以下不属于 MVC 设计模式中 3 个模块的是（　　）。
   A. 模型　　　　　B. 表示层　　　　　C. 视图　　　　　D. 控制器
2. 在 MVC 模式中，（　　）用于客户端应用程序的图形数据表示，与实际数据处理

无关。

    A. 模型   B. 视图   C. 控制器   D. 数据

3. 在 MVC 设计模式中,( )接收用户请求数据。

    A. HTML   B. JSP   C. Servlet   D. 业务类

4. 使用 MVC 模式设计 Web 应用有什么好处?

5. MVC 中的模型是由 Servlet 负责创建,还是由 JSP 页面负责创建?

6. servlet 对象是在服务器端被创建的,还是在用户端被创建的?

7. 什么是转发?什么是重定向?它们有什么区别?

8. 简述 Servlet 的生命周期与运行原理。

9. servlet 对象初始化时是调用 init()方法还是 service()方法?

10. 在 Servlet 中如何获得用户的会话对象?

过滤器

主要内容

(1) 过滤器的概念。
(2) 过滤器的运行原理。
(3) 过滤器的实际应用。

在开发一个网站时,可能有这样的需求:某些页面只希望几个特定的用户浏览。对于这样的访问权限的控制,该如何实现呢?过滤器(Filter)就可以实现上述需求。过滤器位于服务器处理请求之前或服务器响应请求之前。也就是说,它可以过滤浏览器对服务器的请求,也可以过滤服务器对浏览器的响应,如图 8.1 所示。

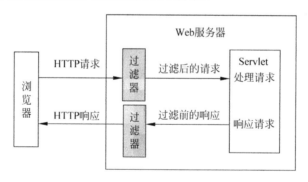

图 8.1 过滤器

8.1 Filter 类与 filter 对象

8.1.1 核心知识

编写一个过滤器类很简单,只要实现 javax.servlet 包中的 Filter 接口。实现 Filter 接口的类习惯地称为一个 Filter 类,这样的类创建的对象又习惯地称为 filter 对象。

8.1.2 能力目标

理解 Filter 类与 filter 对象的概念。

8.1.3 任务驱动

1. 任务的主要内容

新建一个 Web 工程 ch8,在该 Web 工程中编写一个简单的 Filter 类 FirstFilter,Filter 类实现如下功能:不管用户请求该 Web 工程的哪个页面或 Servlet,都会在浏览器中先出现"首先执行过滤器"这样的响应信息。

2. 任务的代码模板

FirstFilter.java 的代码模板如下:

```java
package filter;
import java.io.IOException;
import java.io.PrintWriter;
import javax.servlet.Filter;
import javax.servlet.FilterChain;
import javax.servlet.FilterConfig;
import javax.servlet.ServletException;
import javax.servlet.ServletRequest;
import javax.servlet.ServletResponse;
public class FirstFilter implements 【代码 1】{
    public void destroy() {
    }
    public void doFilter(ServletRequest request,
            ServletResponse response,
            FilterChain chain) throws IOException, ServletException {
        //设置响应类型
        response.setContentType("text/html;charset = GBK");
        //获得输出对象 out
        PrintWriter out = response.getWriter();
        //在浏览器中输出
        out.print("首先执行过滤器<br>");
        //执行下一个过滤器
        chain.doFilter(request, response);
    }
    public void init(FilterConfig fConfig) throws ServletException {
    }
}
```

3. 任务小结或知识扩展

从任务中可以看出,Filter 接口与 Servlet 接口很类似,同样都有 init()与 destroy()方法,还有一个 doFilter()方法类似于 Servlet 接口的 service()方法。下面分别介绍这 3 种方法的功能。

(1) public void init(FilterConfig fConfig) throws ServletException:该方法的功能是初始化过滤器对象。如果为过滤器设置了初始参数,则可以通过 FilterConfig 的 getInitParameter(String paramName)方法获得初始参数值。

(2) public void doFilter（ServletRequestrequest，ServletResponse response，FilterChain chain）throws IOException，ServletException：当 Web 服务器使用 servlet 对象调用 service()方法处理请求前,发现应用了某个过滤器时,Web 服务器就会自动调用该过滤器的 doFilter()方法。在 doFilter()方法中有以下语句：

```
chain.doFilter(request, response);
```

如果执行了该语句,就会执行下一个过滤器,如果没有下一个过滤器,就返回请求目标程序。如果因为某个原因没有执行 chain.doFilter(request，response)；,则请求就不会继续交给以后的过滤器或请求目标程序,这就是拦截请求。

(3) public void destroy()：当 Web 服务器终止服务时,destroy()方法会被执行,使 filter 对象消亡。

4. 代码模板的参考答案

【代码 1】：Filter

8.1.4 实践环节

尝试找一下任务中的 Filter 类编译后的字节码文件。

8.2 filter 对象的部署与运行

8.2.1 核心知识

编写完 Filter 类的源文件,并编译了源文件,这时 Web 服务器是不是就可以运行 filter 对象呢？不可以。需要部署 Filter 后,Web 服务器才可以运行 filter 对象。

与 Servlet 一样,部署过滤器目前有两种方式：一是在 web.xml 中部署 Filter；二是基于注解的方式部署 Filter。

8.2.2 能力目标

掌握部署过滤器的方法。

8.2.3 任务驱动

任务 1：在 web.xml 中部署过滤器

1) 部署 Filter 类

为了在 web.xml 文件里部署 8.1 节中的 FirstFilter,需要在 web.xml 文件里找到 <web-app></web-app>标记,然后在<web-app></web-app>标记中添加如下内容：

```
<filter>
    <filter-name>firstFilter</filter-name>
    <filter-class>filter.FirstFilter</filter-class>
</filter>
```

```
<filter-mapping>
    <filter-name>firstFilter</filter-name>
    <url-pattern>/*</url-pattern>
</filter-mapping>
```

2) 运行 filter 对象

只要用户请求的 URL 和 filter-mapping 的子标记 url-pattern 指定的模式匹配，Web 服务器就会自动调用该 Filter 的 doFilter() 方法。如 8.1 节中的 FirstFilter 过滤器在 web.xml 中的 url-pattern 指定值为 /*,/* 代表任何页面或 Servlet 的请求。

为了测试过滤器，在 Web 工程 ch8 中新建一个 JSP 页面 test.jsp，运行 test.jsp，效果如图 8.2 所示。

图 8.2 首先执行过滤器

3) 有关部署过滤器的标记

① filter 标记及其子标记。web.xml 文件中可以有若干个 filter 标记，该标记的内容由 Web 服务器负责处理。filter 标记中有两个子标记：filter-name 和 filter-class，其中 filter-name 子标记的内容是 Web 服务器创建的 filter 对象的名字。web.xml 文件中可以有若干个 filter 标记，但要求它们的 filter-name 子标记的内容互不相同。filter-class 子标记的内容指定 Web 服务器用哪个类来创建 filter 对象，如果 filter 对象已经创建，那么 Web 服务器就不再使用指定的类创建。

如果在过滤器初始化时，需要读取一些参数的值，则可以在 filter 标记中使用 init-param 子标记设置。例如：

```
<filter>
    <filter-name>firstFilter</filter-name>
    <filter-class>filter.FirstFilter</filter-class>
    <init-param>
        <param-name>encoding</param-name>
        <param-value>GBK</param-value>
    </init-param>
</filter>
```

那么就可以在 Filter 的 init() 方法中，使用参数 fConfig（FilterConfig 的对象）调用 FilterConfig 的 getInitParameter(String paramName) 方法获得参数值。例如：

```
public void init(FilterConfig fConfig) throws ServletException{
    String en = fConfig.getInitParameter("encoding");
}
```

② filter-mapping 标记及其子标记。web.xml 文件中出现一个 filter 标记就会对应地出现一个 filter-mapping 标记。filter-mapping 标记中也有两个子标记：filter-name 和 url-pattern。其中 filter-name 子标记的内容是 Web 服务器创建的 filter 对象的名字（该名字必

须和 filter 标记的子标记 filter-name 的内容相同）；url-pattern 子标记用来指定用户用怎样的模式请求 filter 对象。如果某个 URL 或 Servlet 需应用多个过滤器，则根据 filter-mapping 标记在 web.xml 中出现的先后顺序执行过滤器。

任务 2：基于注解的方式部署过滤器

在 Servlet 3.0 中提供了注解@WebFilter，使不再需要在 web.xml 文件中进行 Filter 的部署描述。但在实际的 Web 工程中，不需要大量开发 Filter，因此，采用任何方式部署过滤器都不太麻烦。

@WebFilter 用于将一个类声明为过滤器，该注解将会在部署时被容器处理，容器将根据具体的属性配置，将相应的类部署为过滤器。该注解的一些常用属性如表 8.1 所示。

表 8.1 @WebFilter 的常用属性

属性名	类型	描述
filterName	String	指定过滤器的 name 属性，等价于 filter-name
value	String[]	该属性等价于 urlPatterns 属性，但两个属性不能同时使用
urlPatterns	String[]	指定一组过滤器的 URL 匹配模式。等价于 url-pattern 标记
servletNames	String[]	指定过滤器将应用于哪些 Servlet。取值是@WebServlet 中的 name 属性的取值，或者是 web.xml 中 servlet-name 的取值
initParams	WebInitParam[]	指定一组过滤器初始化参数，等价于 init-param 标记

表 8.1 中的所有属性均为可选属性，但是 value 或者 urlPatterns 通常是必需的，且两者不能共存，如果同时指定，通常是忽略 value 的取值。

1）任务的主要内容

基于注解的方式部署 Filter——SecondFilter。

2）任务的代码模板

SecondFilter.java 的代码模板如下：

```
package filter;
import java.io.IOException;
import java.io.PrintWriter;
import javax.servlet.Filter;
import javax.servlet.FilterChain;
import javax.servlet.FilterConfig;
import javax.servlet.ServletException;
import javax.servlet.ServletRequest;
import javax.servlet.ServletResponse;
import javax.servlet.annotation.WebFilter;
@WebFilter(filterName = "secondFilter",【代码 1】= { "/*" })
public class SecondFilter implements Filter {
    public void destroy() {
    }
    public void doFilter(ServletRequest request, ServletResponse response,
            FilterChain chain) throws IOException, ServletException {
        //设置响应类型
        response.setContentType("text/html;charset = GBK");
```

```
            //获得输出对象 out
            PrintWriter out = response.getWriter();
            //在浏览器中输出
            out.print("执行第二个过滤器<br>");
            //执行下一个过滤器
            chain.doFilter(request, response);
        }
        public void init(FilterConfig fConfig) throws ServletException {
        }
    }
```

3)任务小结或知识扩展

SecondFilter.java 的代码中使用 @WebFilter(filterName = "secondFilter", urlPatterns = { "/*" })如此部署后,就不必在 web.xml 中部署相应的<filter>和<filter-mapping>元素了,Web 容器会在部署时根据指定的属性将该类发布为 Filter。@WebFilter(filterName = "secondFilter", urlPatterns = { "/*" })等价的 web.xml 部署形式如下:

```xml
<filter>
    <filter-name>secondFilter</filter-name>
    <filter-class>filter.SecondFilter</filter-class>
</filter>
<filter-mapping>
    <filter-name>secondFilter</filter-name>
    <url-pattern>/*</url-pattern>
</filter-mapping>
```

4)代码模板的参考答案

【代码1】: urlPatterns

8.2.4 实践环节

给 8.2.3 节中的 SecondFilter 添加初始化参数,并在该过滤器的 init()方法中获取这些参数。@WebInitParam 注解给过滤器添加参数的示例代码如下:

```
@WebFilter(filterName = "xxxFilter", urlPatterns = { "/*" }, initParams = {
    @WebInitParam(name = "firstParam", value = "one"),
    @WebInitParam(name = "secondParam", value = "two") })
```

8.3 过滤器的应用

8.3.1 核心知识

过滤器是 Servlet 的一种特殊用法,主要用来完成一些通用的操作,比如编码的过滤,判断用户的登录状态,等等。

8.3.2 能力目标

灵活应用过滤器。

8.3.3 任务驱动

任务 1：字符编码过滤器的实现

1）任务的主要内容

编写字符编码过滤器——SetCharacterEncodingFilter。

2）任务的代码模板

SetCharacterEncodingFilter.java 的代码模板如下：

```java
package filter;
import java.io.IOException;
import javax.servlet.Filter;
import javax.servlet.FilterChain;
import javax.servlet.FilterConfig;
import javax.servlet.ServletException;
import javax.servlet.ServletRequest;
import javax.servlet.ServletResponse;
import javax.servlet.annotation.WebFilter;
import javax.servlet.annotation.WebInitParam;
@WebFilter(filterName = "setCharacterEncodingFilter", urlPatterns = { "/*" }, initParams
    = { @WebInitParam(name = "encoding", value = "GBK") })
public class SetCharacterEncodingFilter implements Filter {
    private static String encoding;
    public void destroy() {
    }
    public void doFilter(ServletRequest request, ServletResponse response,
            FilterChain chain) throws IOException, ServletException {
        request.setCharacterEncoding(encoding);
        chain.doFilter(request, response);
    }
    public void init(FilterConfig fConfig) throws ServletException {
        encoding =【代码 1】//获得初始化参数 encoding
    }
}
```

3）任务小结或知识扩展

本书 4.1.3 节介绍了中文乱码的解决方法，其中一种解决方法是，在获取表单信息之前，使用 request 对象调用 setCharacterEncoding(String code)方法设置统一字符编码。使用该方法解决中文乱码问题时，接收参数的每个页面或 Servlet 都需要执行 request.setCharacterEncoding("xxx")语句。为了避免每个页面或 Servlet 都编写 request.setCharacterEncoding("xxx")语句，可以使用过滤器进行字符编码处理。

4）代码模板的参考答案

【代码 1】：fConfig.getInitParameter("encoding");

任务 2：登录验证过滤器的实现

1. 任务的主要内容

新建一个 Web 工程 loginValidateProject，在该 Web 工程中至少编写两个 JSP 页面 login.jsp 与 loginSuccess.jsp，一个 Servlet（由 LoginServlet.java 负责创建）。用户在 login.jsp 页面中输入用户名和密码后，提交给 Servlet，在 Servlet 中判断用户名和密码是否正确，正确则跳转到 loginSuccess.jsp，错误则回到 login.jsp 页面。但该 Web 工程有另外一个要求：除了 login.jsp 页面外，其他页面或 Servlet 都不能直接访问，必须登录成功才能访问。在设计这个 Web 工程时，编写了一个登录验证过滤器并在该 Web 工程中使用。页面运行效果如图 8.3 所示。

(a) 登录页面

(b) 没有登录成功直接运行 loginSuccess.jsp

(c) 登录成功页面

图 8.3 页面运行效果

2. 任务的代码模板

LoginFilter.java（过滤器）的代码如下：

```
package filter;
import java.io.IOException;
import java.io.PrintWriter;
import javax.servlet.Filter;
import javax.servlet.FilterChain;
import javax.servlet.FilterConfig;
import javax.servlet.ServletException;
import javax.servlet.ServletRequest;
import javax.servlet.ServletResponse;
import javax.servlet.annotation.WebFilter;
import javax.servlet.http.HttpServletRequest;
import javax.servlet.http.HttpServletResponse;
```

```java
import javax.servlet.http.HttpSession;
@WebFilter(filterName = "loginFilter", urlPatterns = { "/*" })
public class LoginFilter implements Filter {
    public void destroy() {
    }
    public void doFilter(ServletRequest request, ServletResponse response, FilterChain chain)
    throws IOException, ServletException {
        HttpServletRequest req = (HttpServletRequest) request;
        HttpServletResponse resp = (HttpServletResponse) response;
        HttpSession session = req.getSession(true);
        resp.setContentType("text/html;");
        resp.setCharacterEncoding("UTF-8");
        PrintWriter out = resp.getWriter();
        //得到用户请求的 URI
        String request_uri = req.getRequestURI();
        //得到 Web 应用程序的上下文路径
        String ctxPath = req.getContextPath();
        //去除上下文路径,得到剩余部分的路径
        String uri = request_uri.substring(ctxPath.length());
        //登录页面或 Servlet 不拦截
        if(uri.contains("login.jsp") || uri.contains("loginServlet")){
            chain.doFilter(request, response);
        }else{
            //判断用户是否已经登录
            if (null != session.getAttribute("user")) {
                //执行下一个过滤器
                chain.doFilter(request, response);
            } else {
                out.println("您没有登录,请先登录!3 秒钟后回到登录页面.");
                resp.setHeader("refresh", "3;url=" + ctxPath + "/login.jsp");
                return;
            }
        }
    }
    public void init(FilterConfig fConfig) throws ServletException {
    }
}
```

LoginServlet.java 的代码如下:

```java
package servlet;
import java.io.IOException;
import javax.servlet.ServletException;
import javax.servlet.annotation.WebServlet;
import javax.servlet.http.HttpServlet;
import javax.servlet.http.HttpServletRequest;
import javax.servlet.http.HttpServletResponse;
import javax.servlet.http.HttpSession;
@WebServlet(name = "loginServlet", urlPatterns = { "/loginServlet" })
public class LoginServlet extends HttpServlet {
    private static final long serialVersionUID = 1L;
```

```java
    protected void doGet(HttpServletRequest request,
            HttpServletResponse response) throws ServletException, IOException {
        String username = request.getParameter("name");
        String password = request.getParameter("pwd");
        if(username!=null&&username.equals("filter")){
            if(password!=null&&password.equals("filter")){
                HttpSession session = request.getSession();
                session.setAttribute("user", username);
                response.sendRedirect("loginSuccess.jsp");
            }else{
                response.sendRedirect("login.jsp");
            }
        }else{
            response.sendRedirect("login.jsp");
        }
    }
    protected void doPost(HttpServletRequest request,
            HttpServletResponse response) throws ServletException, IOException {
        doGet(request,response);
    }
}
```

login.jsp 的代码如下：

```jsp
<%@ page language="java" contentType="text/html; charset=GBK" pageEncoding="GBK"%>
<html>
    <head>
        <title>login.jsp</title>
    </head>
    <body bgcolor="lightPink">
        <form action="loginServlet" method="post">
            <table>
                <tr>
                    <td>用户名：</td>
                    <td><input type="text" name="name"/></td>
                </tr>
                <tr>
                    <td>密 码：</td>
                    <td><input type="password" name="pwd"/></td>
                </tr>
                <tr>
                    <td><input type="submit" value="提交"/></td>
                    <td><input type="reset" value="重置"/></td>
                </tr>
            </table>
        </form>
    </body>
</html>
```

loginSuccess.jsp 的代码如下：

```jsp
<%@ page language="java" contentType="text/html; charset=GBK" pageEncoding="GBK"%>
```

```
<html>
    <head>
        <title>loginSuccess.jsp</title>
    </head>
    <body>
        <%
        String username = (String)session.getAttribute("user");
        %>
        恭喜<% = username %>登录成功!
    </body>
</html>
```

3. 任务小结或知识扩展

在 Web 工程中，某些页面或 Servlet 只有用户登录成功才能访问。直接在应用程序每个相关的源代码中判断用户是否登录成功并不是科学的做法。这时可以实现一个登录验证过滤器，不需要在每个相关的源代码中验证用户是否登录成功。

任务中的过滤器要首先检查用户请求的 URL 是不是 login.jsp 或者登录请求 (loginServlet)，这两个值都放在过滤器的初始化参数中。如果用户访问的是 login.jsp 或者登录请求，过滤器就执行 chain.doFilter() 继续请求。如果用户访问的不是 login.jsp 或者登录请求，过滤器先判断用户是否登录成功，登录成功则执行 chain.doFilter() 继续请求，否则重定向到 login.jsp。

8.3.4 实践环节

在任务的 Web 工程 loginValidateProject 中再新建几个 JSP 页面，在没有登录成功的情况下，运行这几个 JSP 页面，看看是什么效果。

8.4 小　　结

本章讲解了 Filter 的概念、原理以及实际应用。Filter 使 Servlet 开发者能够在请求到达 Servlet 之前拦截请求，在 Servlet 处理请求之后修改响应。

习　题　8

1. 简述过滤器的运行原理。
2. Filter 接口中有哪些方法？它们分别具有什么功能？
3. 在 web.xml 中部署过滤器需要哪些标记？这些标记的作用是什么？

EL 与 JSTL

主要内容

(1) 表达式语言(EL)。
(2) JSP 标准标签库(JSTL)。

在 JSP 页面中可以使用 Java 代码来实现页面显示逻辑,但网页中夹杂着 HTML 与 Java 代码,给网页的设计与维护带来困难。可以使用 EL 来访问和处理应用程序的数据,也可以使用 JSTL 来替换网页中实现页面显示逻辑的 Java 代码。这样 JSP 页面就尽量减少了 Java 代码的使用,为以后的维护提供了方便。

本章涉及的 Java 源文件保存在工程 ch9 的 src 目录中,涉及的 JSP 页面保存在工程 ch9 的 WebContent 目录中。

9.1 表达式语言 EL

EL 是 JSP 2.0 规范中增加的,它的基本语法如下:

$﹛表达式﹜

EL 表达式类似于 JSP 表达式＜%＝表达式%＞,EL 语句中的表达式值会被直接送到浏览器显示。通过 page 指令的 isELIgnored 属性来说明是否支持 EL 表达式。isELIgnored 属性值为 false 时,JSP 页面可以使用 EL 表达式;isELIgnored 属性值为 true 时,JSP 页面不能使用 EL 表达式。isELIgnored 属性值默认为 false。

9.1.1 核心知识

1. EL 基本语法

EL 的语法简单,使用方便。它以 ${ 开始,} 结束。

1) [ ]与.运算符

EL 使用[ ]和.运算符来访问数据,主要使用 EL 获取对象的属性,包括获取 JavaBean 的属性值,获取数组中的元素以及获取集合对象中的元素。对于 null 值直接以空字符串显示,而不是 null,运算时也不会发生错误或空指针异常。所以在使用 EL 访问对象的属性时，不需要判断对象是否为 null 对象。这样就为编写程序提供了方便。

① 获取 JavaBean 的属性值。假设在 JSP 页面中有以下代码：

```
<jsp:getProperty property = "age" name = "user"/>
```

那么，可以使用 EL 获取 user 的属性 age，代码如下：

```
${user.age}
```

或

```
${user["age"]}
```

其中，点运算符前面为 JavaBean 的对象 user，后面为该对象的属性 age，表示利用 user 对象的 getAge()方法取值并显示在网页上。

② 获取数组中的元素。假设在 JSP 页面或 Servlet 中有以下代码：

```
<%
    String dogs[] = {"lili","huahua","guoguo"};
    request.setAttribute("array", dogs);
%>
```

那么，在 JSP 中可以使用 EL 取出数组中的元素，代码如下：

```
${array[0]}
${array[1]}
${array[2]}
```

③ 获取集合对象中的元素。假设在 JSP 页面或 Servlet 中有以下代码：

```
<%
    ArrayList<UserBean> users = new ArrayList<UserBean>();
    UserBean ub1 = new UserBean("zhang",20);
    UserBean ub2 = new UserBean("zhao",50);
    users.add(ub1);
    users.add(ub2);
    request.setAttribute("array", users);
%>
```

其中，UserBean 有两个属性：name 和 age，那么在 JSP 页面中可以使用 EL 取出 UserBean 中的属性，代码如下：

```
${array[0].name}   ${array[0].age}
${array[1].name}   ${array[1].age}
```

2）算术运算符

在 EL 表达式中有 5 个算术运算符，如表 9.1 所示。

表 9.1　EL 的算术运算符

算术运算符	说　明	示　例	结　果
＋	加	${13＋2}	15
－	减	${13－2}	11
＊	乘	${13＊2}	26
/(或 div)	除	${13/2}或${13 div 2}	6.5
%(或 mod)	取模(求余)	${13%2}或${13 mod 2}	1

3）关系运算符

在 EL 表达式中有 6 个关系运算符,如表 9.2 所示。

表 9.2　EL 的关系运算符

关系运算符	说　明	示　例	结　果
＝＝(或 eq)	等于	${13＝＝2}或${13 eq 2}	false
!＝(或 ne)	不等于	${13!＝2}或${13 ne 2}	true
＜(或 lt)	小于	${13＜2}或${13 lt 2}	false
＞(或 gt)	大于	${13＞2}或${13 gt 2}	true
＜＝(或 le)	小于等于	${13＜＝2}或${13 le 2}	false
＞＝(或 ge)	大于等于	${13＞＝2}或${13 ge 2}	true

4）逻辑运算符

在 EL 表达式中有 3 个逻辑运算符,如表 9.3 所示。

表 9.3　EL 的逻辑运算符

逻辑运算符	说　明	示　例	结　果
&&(或 and)	逻辑与	如果 A 为 true,B 为 false,则 A && B(或 A and B)	false
||(或 or)	逻辑或	如果 A 为 true,B 为 false,则 A || B(或 A or B)	true
!(或 not)	逻辑非	如果 A 为 true,则! A(或 not A)	false

5）empty 运算符

empty 运算符用于检测一个值是否为 null,例如,变量 A 不存在,则 ${empty A}返回的结果为 true。

6）条件运算符

EL 中的条件运算符是"？:",例如,${A？B:C},如果 A 为 true,计算 B 并返回其结果,如果 A 为 false,计算 C 并返回其结果。

2. EL 隐含对象

EL 隐含对象共有 11 个,这里只介绍几个常用的 EL 隐含对象:即 pageScope、requestScope、sessionScope、applicationScope、param 以及 paramValues。

1）与作用范围相关的隐含对象

与作用范围有关的 EL 隐含对象有 pageScope、requestScope、sessionScope 和 applicationScope,分别可以获取 JSP 隐含对象 pageContext、request、session 和 application 中的数据。如果在 EL 中没有使用隐含对象指定作用范围,则会依次从 page、request、 session、application 范围查找,找到就直接返回,不再继续查找下去,如果所有范围都没有找到,就返回空字符串。获取数据的格式如下:

    ${EL 隐含对象.关键字对象.属性}

或

    ${EL 隐含对象.关键字对象}

例如:

```
<jsp:useBean id = "user" class = "bean.UserBean" scope = "page"/>
<jsp:setProperty name = "user" property = "name" value = "EL 隐含对象" />
name: ${pageScope.user.name}
```

再如,在 JSP 页面或 Servlet 中有这样一段话:

```
<%
    ArrayList<UserBean> users = new ArrayList<UserBean>();
    UserBean ub1 = new UserBean("zhang",20);
    UserBean ub2 = new UserBean("zhao",50);
    users.add(ub1);
    users.add(ub2);
    request.setAttribute("array", users);
%>
```

其中,UserBean 有两个属性:即 name 和 age,那么在 request 有效的范围内可以使用 EL 取出 UserBean 的属性,代码如下:

```
${requestScope.array[0].name}    ${requestScope.array[0].age}
${requestScope.array[1].name}    ${requestScope.array[1].age}
```

2) 与请求参数相关的隐含对象

与请求参数相关的 EL 隐含对象有 param 和 paramValues。获取数据的格式如下:

    ${EL 隐含对象.参数名}

比如,input.jsp 的代码如下:

```
<form method = "post" action = "param.jsp">
    <p>姓名:<input type = "text" name = "username" size = "15" /></p>
    <p>兴趣:
    <input type = "checkbox" name = "habit" value = "看书"/>看书
    <input type = "checkbox" name = "habit" value = "玩游戏"/>玩游戏
    <input type = "checkbox" name = "habit" value = "旅游"/>旅游
    <p>
    <input type = "submit" value = "提交"/>
</form>
```

那么,在 param.jsp 页面中可以使用 EL 获取参数值,代码如下:

```
<% request.setCharacterEncoding("GBK"); %>
<body>
```

```
<h2>EL 隐含对象 param、paramValues </h2>
姓名：${param.username}</br>
兴趣：
${paramValues.habit[0]}
${paramValues.habit[1]}
${paramValues.habit[2]}
```

9.1.2 能力目标

能够灵活使用表达式语言 EL。

9.1.3 任务驱动

1. 任务的主要内容

编写一个 Servlet，在该 Servlet 中使用 request 对象存储数据，然后从该 Servlet 转发到 show.jsp 页面，在 show.jsp 页面中显示 request 对象的数据。首先，运行 Servlet，在 IE 地址栏中输入

```
http://localhost:8080/ch9/saveServlet
```

程序运行结果如图 9.1 所示。

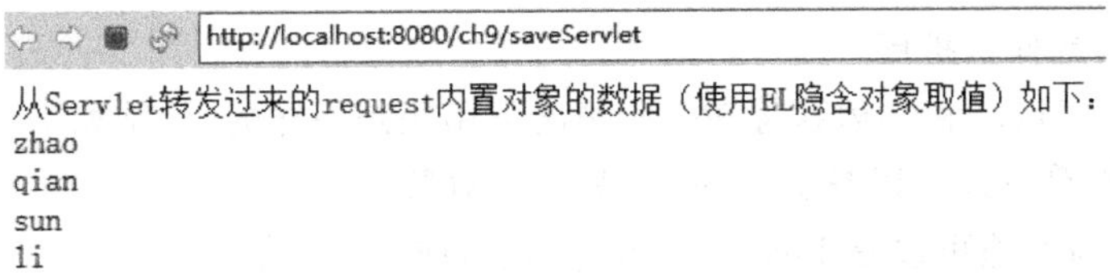

图 9.1 使用 EL 内置对象获取 JSP 内置对象的数据

2. 任务的代码模板

SaveServlet.java 的代码如下：

```java
package servlet;
import java.io.IOException;
import javax.servlet.RequestDispatcher;
import javax.servlet.ServletException;
import javax.servlet.annotation.WebServlet;
import javax.servlet.http.HttpServlet;
import javax.servlet.http.HttpServletRequest;
import javax.servlet.http.HttpServletResponse;
@WebServlet(name = "saveServlet", urlPatterns = { "/saveServlet" })
public class SaveServlet extends HttpServlet {
    private static final long serialVersionUID = 1L;
    protected void doGet(HttpServletRequest request,
            HttpServletResponse response) throws ServletException, IOException {
        doPost(request, response);
    }
    protected void doPost(HttpServletRequest request,
            HttpServletResponse response) throws ServletException, IOException {
```

```java
            String names[] = { "zhao", "qian", "sun", "li" };
            request.setAttribute("name", names);
            RequestDispatcher dis = request.getRequestDispatcher("show.jsp");
            dis.forward(request, response);
        }
    }
```

show.jsp 的代码模板如下：

```jsp
<%@ page language="java" contentType="text/html; charset=GBK" pageEncoding="GBK" %>
<html>
    <head>
        <title>EL 内置对象</title>
    </head>
    <body>
        从 Servlet 转发过来的 request 内置对象的数据(使用 EL 隐含对象取值)如下：<br>
            【代码 1】<br>
            【代码 2】<br>
            【代码 3】<br>
            【代码 4】<br>
    </body>
</html>
```

3. 任务小结或知识扩展

EL 中点运算符(.)和[]运算符在一些情况下用法是一样的，总结如下。

(1) (.)运算符左边可以是 JavaBean 或 map 对象。

(2) []运算符左边可以是 JavaBean、map、数组或 list 对象。

使用 EL 如何取得 map 对象中的值呢？假设在 JSP 页面中有以下代码：

```jsp
<%
    HashMap<String,String> map = new HashMap<String,String>();
    map.put("fisrt", "第一");
    map.put("second", "第二");
    request.setAttribute("number", map);
%>
```

那么在页面某处可以使用 EL 获得 map 中的值，代码如下：

```
${number.fisrt}
${number.second}
```

或

```
${number["fisrt"]}
${number["second"]}
```

4. 代码模板的参考答案

【代码 1】：${requestScope.name[0]}
【代码 2】：${requestScope.name[1]}
【代码 3】：${requestScope.name[2]}

【代码 4】：${requestScope.name[3]}

9.1.4 实践环节

将本节任务中 Servlet 的 doPost()方法修改如下：

```
protected void doPost(HttpServletRequest request,
    HttpServletResponse response) throws ServletException, IOException {
//String names[] = { "zhao", "qian", "sun", "li" };
Map<String, String> names = new HashMap<String, String>();
names.put("first", "zhao");
names.put("second", "qian");
names.put("third", "sun");
names.put("forth", "li");
request.setAttribute("name", names);
RequestDispatcher dis = request.getRequestDispatcher("show.jsp");
dis.forward(request, response);
}
```

请修改 show.jsp，显示 map 中的数据。

9.2 JSP 标准标签库 JSTL

JSTL 规范由 Sun 公司制定，Apache 的 Jakarta 小组负责实现。在写作本书时，JSTL 的最新版本是 JSTL 1.2。JSTL 标准标签库由 5 个不同功能的标签库组成，包括 Core、I18N、XML、SQL 以及 Functions，本书只简要介绍 JSTL 的 Core 标签库中几个常用标签。

9.2.1 核心知识

1. 配置 JSTL

JSTL 1.2 现在已经是 Java EE5 的一个组成部分，如果采用支持 Java EE5 或 Java EE6 的集成开发环境开发 Web 应用程序，就不需要再配置 JSTL。但本书采用的是 Eclipse 平台，因此需要配置 JSTL。配置 JSTL 的步骤如下。

1）复制 JSTL 的标准实现

在 Tomcat 的\webapps\examples\WEB-INF\lib 目录下，找到 jstl.jar 和 standard.jar 文件，然后复制到 Web 工程的 WEB-INF\lib 目录下。

2）使用 taglib 标记定义前缀与 uri 引用

在 JSP 页面中使用 taglib 标记定义前缀与 uri 引用，代码如下：

```
<%@ taglib prefix="c" uri="http://java.sun.com/jsp/jstl/core" %>
```

2. 核心标签库的通用标签

1）c:out 标签

c:out 用来显示数据的内容，与<%=表达式%>或${表达式}类似。格式如下：

```
<c:out value="输出的内容" [default="defaultValue"]/>
```

或

```
<c:out value = "输出的内容">
    defaultValue
</c:out>
```

其中,value 值可以是一个 EL 表达式,也可以是一个字符串;default 可有可无,当 value 值不存在时,就输出 defaultValue。例如:

```
<c:out value = "${param.data}" default = "没有数据" />
<br>
<c:out value = "${param.nothing}" />
<br>
<c:out value = "这是一个字符串" />
```

输出的结果如图 9.2 所示。

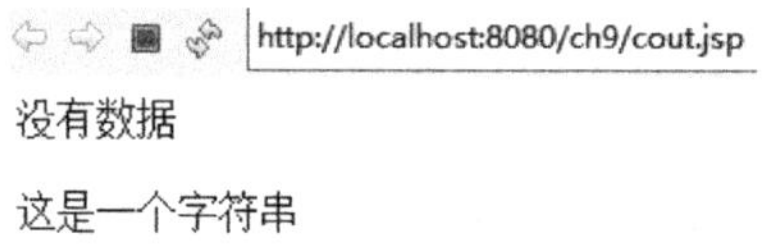

```
⇦ ⇨ ■ ⌕  http://localhost:8080/ch9/cout.jsp
没有数据
这是一个字符串
```

图 9.2 c:out 标签

2) c:set 标签

① 设置作用域变量。可以使用 c:set 在 page、request、session、application 等范围内设置一个变量。格式如下:

```
<c:set value = "value" var = "varName" [scope = "page|request|session|application"]/>
```

将 value 值赋值给变量 varName。例如:

```
<c:set value = "zhao" var = "userName" scope = "session"/>
```

相当于

```
<% session.setAttribute("userName","zhao"); %>
```

② 设置 JavaBean 的属性。使用 c:set 设置 JavaBean 的属性时,必须使用 target 属性进行设置。格式如下:

```
<c:set value = "value" target = "target" property = "propertyName"/>
```

将 value 赋值给 target 对象(JaveBean 对象)的 propertyName 属性。如果 target 为 null 或没有 set 方法则抛出异常。

3) c:remove 标签

如果要删除某个变量,则可以使用 c:remove 标签。例如:

```
<c:remove var = "userName" scope = "session"/>
```

相当于

```
<% session.removeAttribute("userName") %>
```

3. 核心标签库的流程控制标签

1) c:if 标签

c:if 标签实现 if 语句的作用,具体语法格式如下:

```
<c:if test = "条件表达式">
    主体内容
</c:if>
```

其中,条件表达式可以是 EL 表达式,也可以是 JSP 表达式。如果表达式的值为 true,则会执行 c:if 的主体内容,但是没有相对应的 c:else 标签。如果想在条件成立时执行一块内容,不成立时执行另一块内容,则可以使用 c:choose、c:when 及 c:otherwise 标签。

2) c:choose、c:when 及 c:otherwise 标签

c:choose、c:when 及 c:otherwise 标签实现 if/elseif/else 语句的作用。具体语法格式如下:

```
<c:choose>
    <c:when test = "条件表达式 1">
        主体内容 1
    </c:when>
    <c:when test = "条件表达式 2">
        主体内容 2
    </c:when>
    <c:otherwise>
        表达式都不正确时,执行的主体内容
    </c:otherwise>
</c:choose>
```

【例 9.1】 编写一个 JSP 页面 ifelse.jsp,在该页面中使用 c:set 标签把两个字符串设置为 request 范围内的变量。使用 c:if 标签求出这两个字符串的最大值(按字典顺序比较大小),使用 c:choose、c:when 及 c:otherwise 标签求出这两个字符串的最小值。

例 9.1 页面文件 ifelse.jsp 的代码如下:

```
<%@ page language = "java" contentType = "text/html; charset = GBK" pageEncoding = "GBK" %>
<%@ taglib prefix = "c" uri = "http://java.sun.com/jsp/jstl/core" %>
<html>
    <head>
        <title>ifelse.jsp</title>
    </head>
    <body>
        <c:set value = "if" var = "firstNumber" scope = "request" />
        <c:set value = "else" var = "secondNumber" scope = "request" />
        <c:if test = "${firstNumber > secondNumber}">
            最大值为 ${firstNumber}
        </c:if>
        <c:if test = "${firstNumber < secondNumber}">
            最大值为 ${secondNumber}
        </c:if>
        <c:choose>
```

```
            <c:when test="${firstNumber<secondNumber}">
                最小值为${firstNumber}
            </c:when>
            <c:otherwise>
                最小值为${secondNumber}
            </c:otherwise>
        </c:choose>
    </body>
</html>
```

c:when 及 c:otherwise 必须放在 c:choose 中。当 c:when 的 test 结果为 true 时,会输出 c:when 的主体内容,而不理会 c:otherwise 的内容。c:choose 中可有多个 c:when,程序会从上到下进行条件判断,如果有一个 c:when 的 test 结果为 true,就输出其主体内容,之后的 c:when 就不再执行。如果所有的 c:when 的 test 结果都为 false,则会输出 c:otherwise 的内容。c:if 与 c:choose 也可以嵌套使用,例如:

```
<c:set value="fda" var="firstNumber" scope="request"/>
<c:set value="else" var="secondNumber" scope="request"/>
<c:set value="ddd" var="threeNumber" scope="request"/>
<c:if test="${firstNumber>secondNumber}">
    <c:choose>
        <c:when test="${firstNumber<threeNumber}">
            最大值为${threeNumber}
        </c:when>
        <c:otherwise>
            最大值为${firstNumber}
        </c:otherwise>
    </c:choose>
</c:if>
<c:if test="${secondNumber>firstNumber}">
    <c:choose>
        <c:when test="${secondNumber<threeNumber}">
            最大值为${threeNumber}
        </c:when>
        <c:otherwise>
            最大值为${secondNumber}
        </c:otherwise>
    </c:choose>
</c:if>
```

4. 核心标签库的迭代标签

1) c:forEach 标签

c:forEach 标签可以实现程序中的 for 循环。语法格式如下:

```
<c:forEach var="变量名" items="数组或collection对象">
    循环体
</c:forEach>
```

其中，items 属性可以是数组或 collection 对象，每次循环读取对象中的一个元素，并赋值给 var 属性指定的变量，之后就可以在循环体使用 var 指定的变量获取对象的元素。例如，在 JSP 页面或 Servlet 中有以下代码：

```
<%
    ArrayList<UserBean> users = new ArrayList<UserBean>();
    UserBean ub1 = new UserBean("zhao",20);
    UserBean ub2 = new UserBean("qian",40);
    UserBean ub3 = new UserBean("sun",60);
    UserBean ub4 = new UserBean("li",80);
    users.add(ub1);
    users.add(ub2);
    users.add(ub3);
    users.add(ub4);
    request.setAttribute("usersKey", users);
%>
```

那么，在 JSP 页面中可以使用 c:forEach 循环遍历出数组中的元素。代码如下：

```
<table>
    <tr>
        <th>姓名</th>
        <th>年龄</th>
    </tr>
    <c:forEach var="user" items="${requestScope.usersKey}">
        <tr>
            <td>${user.name}</td>
            <td>${user.age}</td>
        </tr>
    </c:forEach>
</table>
```

有些情况下，需要为 c:forEach 标签指定 begin、end、step 和 varStatus 属性。begin 为迭代时的开始位置，默认值为 0；end 为迭代时的结束位置，默认值是最后一个元素；step 为迭代步长，默认值为 1；varStatus 代表迭代变量的状态，包括 count（迭代的次数）、index（当前迭代的索引，第一个索引为 0）、first（是否是第一个迭代对象）和 last（是否是最后一个迭代对象）。例如：

```
<table border=1>
    <tr>
        <th>Value</th>
        <th>Square</th>
        <th>Index</th>
    </tr>
    <c:forEach var="x" varStatus="status" begin="0" end="10" step="2">
        <tr>
            <td>${x}</td>
            <td>${x * x}</td>
```

```
            <td>${status.index}</td>
        </tr>
    </c:forEach>
</table>
```

上述程序运行结果如图 9.3 所示。

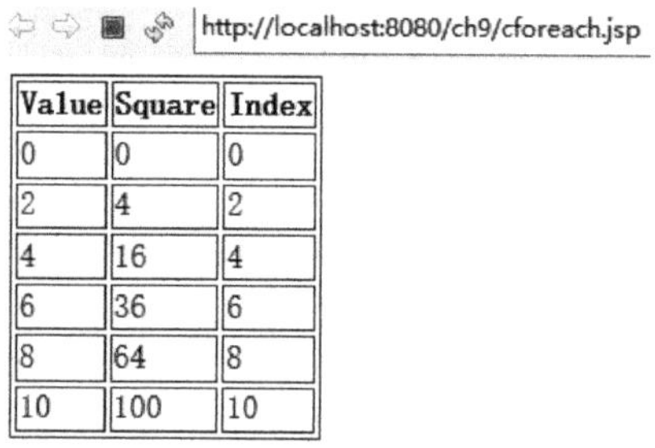

Value	Square	Index
0	0	0
2	4	2
4	16	4
6	36	6
8	64	8
10	100	10

图 9.3　c:forEach 标签

2) c:forTokens 标签

c:forTokens 用于迭代字符串中由分隔符分隔的各成员，它通过 java. util. StringTokenizer 实例来完成字符串的分隔，属性 items 和 delims 作为构造 StringTokenizer 实例的参数。语法格式如下：

```
<c:forTokens var="变量名" items="要迭代的 String 对象" delims="指定分隔字符串的分隔
符">
    循环体
</c:forTokens>
```

例如：

```
<c:forTokens items="chenheng1:chenheng2:chenheng3" delims=":" var="name">
    ${name}<br>
</c:forTokens>
```

上述程序运行结果如图 9.4 所示。

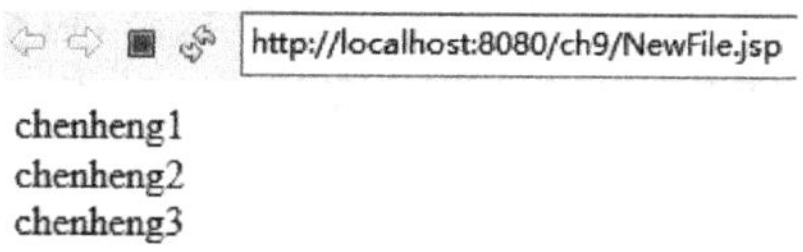

图 9.4　c:forTokens 标签

c:forTokens 标签与 c:forEach 标签一样，也有 begin、end、step 和 varStatus 属性，并且用法一样，这里不再赘述。

9.2.2 能力目标

能够灵活使用 JSTL 核心标签库。

9.2.3 任务驱动

1. 任务的主要内容

首先,在 outEven.jsp 页面中有以下代码:

```
<%
    ArrayList<UserBean> users = new ArrayList<UserBean>();
    UserBean ub1 = new UserBean("zhao",20);
    UserBean ub2 = new UserBean("qian",40);
    UserBean ub3 = new UserBean("sun",60);
    UserBean ub4 = new UserBean("li",80);
    users.add(ub1);
    users.add(ub2);
    users.add(ub3);
    users.add(ub4);
    request.setAttribute("usersKey", users);
%>
```

其次,在该JSP页面中使用 c:forEach 循环遍历出数组 users 中下标为偶数的元素(即 zhao、sun)。页面运行效果如图 9.5 所示。

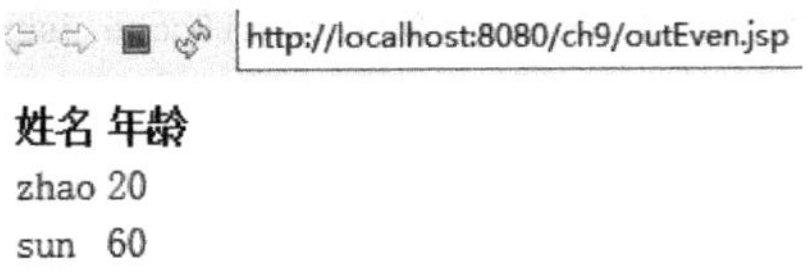

图 9.5 使用 c:forEach 输出下标为偶数的元素

2. 任务的代码模板

outEven.jsp 的代码模板如下:

```
<%@ page language="java" contentType="text/html; charset=GBK" pageEncoding="GBK"%>
<%@ taglib prefix="c" uri="http://java.sun.com/jsp/jstl/core"%>
<%@ page import="bean.UserBean"%>
<%@ page import="java.util.*"%>
<html>
    <head>
        <title>outEven.jsp</title>
    </head>
    <body>
        <%
            ArrayList<UserBean> users = new ArrayList<UserBean>();
            UserBean ub1 = new UserBean("zhao",20);
            UserBean ub2 = new UserBean("qian",40);
            UserBean ub3 = new UserBean("sun",60);
            UserBean ub4 = new UserBean("li",80);
            users.add(ub1);
            users.add(ub2);
            users.add(ub3);
            users.add(ub4);
            request.setAttribute("usersKey", users);
```

```
        %>
        <table>
            <tr>
                <th>姓名</th>
                <th>年龄</th>
            </tr>
            【代码 1】<! -- 使用 forEach 标签 -->
            <tr>
                <td>${user.name}</td>
                <td>${user.age}</td>
            </tr>
            【代码 2】
        </table>
    </body>
</html>
```

3. 任务小结或知识扩展

c:forEach 标签是 JSTL 中非常重要的一个迭代标签，它类似于 Java 语言的 for 循环语句。

4. 代码模板的参考答案

【代码 1】: `<c:forEach var="user" items="${requestScope.usersKey}" step="2">`
【代码 2】: `</c:forEach>`

9.2.4 实践环节

编写一个 JSP 页面，在该页面中使用 c:forEach 标签输出九九乘法表。页面运行效果如图 9.6 所示。

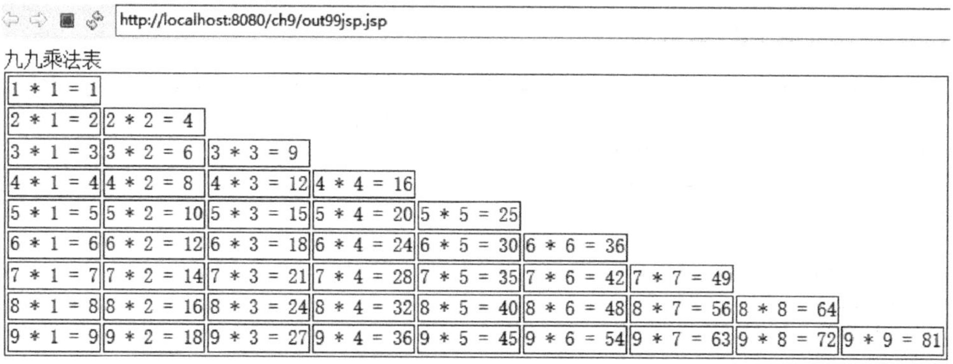

1 * 1 = 1								
2 * 1 = 2	2 * 2 = 4							
3 * 1 = 3	3 * 2 = 6	3 * 3 = 9						
4 * 1 = 4	4 * 2 = 8	4 * 3 = 12	4 * 4 = 16					
5 * 1 = 5	5 * 2 = 10	5 * 3 = 15	5 * 4 = 20	5 * 5 = 25				
6 * 1 = 6	6 * 2 = 12	6 * 3 = 18	6 * 4 = 24	6 * 5 = 30	6 * 6 = 36			
7 * 1 = 7	7 * 2 = 14	7 * 3 = 21	7 * 4 = 28	7 * 5 = 35	7 * 6 = 42	7 * 7 = 49		
8 * 1 = 8	8 * 2 = 16	8 * 3 = 24	8 * 4 = 32	8 * 5 = 40	8 * 6 = 48	8 * 7 = 56	8 * 8 = 64	
9 * 1 = 9	9 * 2 = 18	9 * 3 = 27	9 * 4 = 36	9 * 5 = 45	9 * 6 = 54	9 * 7 = 63	9 * 8 = 72	9 * 9 = 81

图 9.6 使用 c:forEach 打印九九乘法表

9.3 小　　结

本章重点介绍了 EL 表达式和 JSTL 核心标签库的用法。EL 与 JSTL 的应用大大提高了编程效率，并且降低了维护难度。

习 题 9

1. 在 Web 应用程序中有以下程序代码段,执行后转发到某个 JSP 页面:

```
ArrayList<String> dogNames = new ArrayList<String>();
dogNames.add("goodDog");
request.setAttribute("dogs", dogNames);
```

以下( )选项可以正确地使用 EL 取得数组中的值。

  A. ${dogs.0}

  B. ${dogs[0]}

  C. ${dogs.[0]}

  D. ${dogs "0"}

2. ( )JSTL 标签可以实现 Java 程序中的 if 语句功能。

  A. <c:set>

  B. <c:out>

  C. <c:forEach>

  D. <c:if>

3. ( )不是 EL 的隐含对象。

  A. request

  B. pageScope

  C. sessionScope

  D. applicationScope

4. ( )JSTL 标签可以实现 Java 程序中的 for 语句功能。

  A. <c:set>

  B. <c:out>

  C. <c:forEach>

  D. <c:if>

第 10 章

文件的上传与下载

主要内容

(1) 基于 Servlet 3.0 的文件上传。
(2) 文件的下载。

在 JSP 应用程序中实现文件上传有两种常见的组件：jspsmart 组件和 commons-fileupload 组件，但在本章介绍的是基于 Servlet 3.0 的文件上传。

本章涉及的 Java 源文件保存在工程 ch10 的 src 中，涉及的 JSP 页面保存在工程 ch10 的 WebContent 中。

10.1 基于 Servlet 3.0 的文件上传

文件的上传与下载在 Web 应用程序中是很常见的功能。例如，在 OA 办公系统中，用户可以使用文件上传来提交公文。

1. 基于表单的文件上传

表单元素：<input type="file">，在浏览器中会显示一个输入框和一个按钮，输入框可供用户填写本地文件的文件名和路径名，按钮可以让浏览器打开一个文件选择框供用户选择文件。

文件上传的表单例子如下：

```
<form action="upload" method="post" enctype="multipart/form-data">
    <input type="file" name="headImage"><br>
    <input type="submit" value="上传">
</form>
```

基于表单的文件上传，不要忘记使用 enctype 属性，并将它的值设置为 multipart/form-data。同时，表单的提交方式设置为 post。

2. 基于 Servlet 3.0 的文件上传

Servlet 3.0 之前的版本不能直接支持文件上传，需要使用第三方框架来实现，而且使用起来也不够简单。Servlet 3.0 已经提供了这个功能,而且使用也非常简单。

让 Servlet 支持文件上传,需要做以下两件事情。

① 给 Servlet 添加@MultipartConfig 注解。

② 从 HttpServletRequest 对象中获取 Part 文件对象。

1) @MultipartConfig 注解

@MultipartConfig 注解主要是为了辅助 Servlet 3.0 中 HttpServletRequest 提供对上传文件的支持。该注解标注在 Servlet 上面,以表示该 Servlet 希望处理的请求的 MIME 类型是 multipart/form-data。另外,它还提供了若干属性用于简化对上传文件的处理。该注解的常用属性如表 10.1 所示。

表 10.1 @MultipartConfig 常用属性

属性名	类型	是否可选	描 述
location	String	是	指定上传文件存放的目录。当指定了 location 后,在调用 Part 的 write(String fileName)方法把文件写入磁盘时,文件名称可以不用带路径,但是如果 fileName 带了绝对路径,那将以 fileName 所带路径为准把文件写入磁盘
maxFileSize	long	是	指定上传文件的最大值,单位是字节。默认值为−1,表示没有限制
maxRequestSize	long	是	指定上传文件的个数,应用在多文件上传。默认值为−1,表示没有限制

注意:即使没有使用@MultipartConfig 注解设置属性,也要把该注解加到 Servlet 的上面。

2) Part 接口

每一个文件用一个 javax.servlet.http.Part 对象来表示。单个文件上传时,在 Servlet 中可以通过 HttpServletRequest 的对象 request 调用方法 getPart(String name)获得 Part 文件对象。其中,参数 name 为文件域的名称。例如:

```
Part photo = request.getPart("resPath");
```

多文件上传时,在 Servlet 中可以通过 HttpServletRequest 的对象 request 调用方法 getParts()获得 Part 文件对象集合。例如:

```
Collection<Part> photos = request.getParts();
```

Part 接口的常用方法如表 10.2 所示。

表 10.2 Part 接口的常用方法

序号	方 法	功 能 说 明
1	void delete()	删除任何相关的临时磁盘文件
2	String getContentType()	获得客户端浏览器设置的文件数据项的 MIME 类型
3	String getHeader(String name)	获得指定的 part 头的一个字符串。例如,getHeader("content-disposition") 返回 form-data; name = " xxx"; filename="xxx"

续表

序号	方　　法	功　能　说　明
4	InputStream getInputStream( )	获得一个输入流,通过这个输入流来读取文件的内容
5	String getName( )	获得表单文件域的名称
6	long getSize( )	获得文件的大小
7	void write(String fileName)	将文件上传到 fileName 指定的目录里

Servlet 3.0 规范中不提供获取上传文件名的方法,但可以通过 Part 对象调用 getHeader("content-disposition")方法间接获取。根据 Part 对象获取文件名的方法 getFileName(Part part)在本章工具类 MyUtil 中。

本章工具类 MyUtil.java 的代码如下:

```java
package util;
import java.io.UnsupportedEncodingException;
import java.text.SimpleDateFormat;
import java.util.Date;
import javax.servlet.http.Part;
public class MyUtil {
    //主键产生器
    public static String getStringID(){
        String id = null;
        Date date = new Date();
        SimpleDateFormat sdf = new SimpleDateFormat("yyyyMMddHHmmssSSS");
        id = sdf.format(date);
        return id;
    }
    //中文文件名字符编码转换方法
    public static String toUTF8String(String str){
        StringBuffer sb = new StringBuffer();
        int len = str.length();
        for(int i = 0; i < len; i++){
            //取出字符中的每个字符
            char c = str.charAt(i);
            //Unicode 码值在 0～255,不作处理
            if(c >= 0 && c <= 255){
                sb.append(c);
            }else{//转换 UTF-8 编码
                byte b[];
                try {
                    b = Character.toString(c).getBytes("UTF-8");
                } catch (UnsupportedEncodingException e) {
                    e.printStackTrace();
                    b = null;
                }
                //转换为 %HH 的字符串形式
                for(int j = 0; j < b.length; j ++){
                    int k = b[j];
                    if(k < 0){
```

```
                k & = 255;
            }
            sb.append("%" + Integer.toHexString(k).toUpperCase());
        }
    }
}
    return sb.toString();
}
//从 Part 中获得原始文件名
public static String getFileName(Part part){
    if(part == null)
        return null;
    //fileName 形式为: form - data; name = "resPath"; filename = "20140920_110531.jpg"
    String fileName = part.getHeader("content-disposition");
    //没有选择文件
    if(fileName.lastIndexOf("=") + 2 == fileName.length() - 1)
        return null;
    return fileName.substring(fileName.lastIndexOf("=") + 2, fileName.length() - 1);
}
}
```

10.1.2 能力目标

掌握基于 Servlet 3.0 的文件上传方法。

10.1.3 任务驱动

任务 1：单个文件上传

1) 任务的主要内容

编写一个页面 uploadHttpOne.jsp，在页面中选择文件提交给 urlPatterns 为{"/uploadHttpOneServet"}的 servlet 对象（由 UploadHttpOneServet 类负责创建）。在 Servlet 中进行单个文件上传。页面 uploadHttpOne.jsp 的运行效果如图 10.1 所示。

图 10.1 uploadHttpOne.jsp 的运行效果

2) 任务的代码模板

页面文件 uploadHttpOne.jsp 的代码模板如下：

```
<%@ page language="java" contentType="text/html; charset=GBK" pageEncoding="GBK"%>
<html>
    <head>
        <title>HttpServletRequest 对文件上传的支持</title>
    </head>
```

```html
<body>
    <form action = "uploadHttpOneServet" method = "post"【代码1】>
        <table>
            <tr>
                <td>文件描述:</td>
                <td><input type = "text" name = "filediscription"></td>
            </tr>
            <tr>
                <td>请选择文件:</td>
                <td><input【代码2】name = "resPath"></td>
            </tr>
            <tr>
                <td align = "right"><input type = "reset" value = "重填"></td>
                <td><input type = "submit" value = "上传"></td>
            </tr>
        </table>
    </form>
</body>
</html>
```

uploadHttpOneServet.java 的代码模板如下:

```java
package servlet;
import java.io.File;
import java.io.IOException;
import java.io.PrintWriter;
import javax.servlet.ServletException;
import javax.servlet.annotation.MultipartConfig;
import javax.servlet.annotation.WebServlet;
import javax.servlet.http.HttpServlet;
import javax.servlet.http.HttpServletRequest;
import javax.servlet.http.HttpServletResponse;
import javax.servlet.http.Part;
import util.MyUtil;
@WebServlet(name = "uploadHttpOneServet", urlPatterns = { "/uploadHttpOneServet" })
@MultipartConfig(maxFileSize = 10 * 1024 * 1024)//设置上传文件的最大值为10MB
public class UploadHttpOneServet extends HttpServlet {
    private static final long serialVersionUID = 1L;
    protected void doGet(HttpServletRequest request, HttpServletResponse response) throws ServletException, IOException {
        doPost(request, response);
    }
    protected void doPost(HttpServletRequest request, HttpServletResponse response) throws ServletException, IOException {
        //设置响应的内容类型
        response.setContentType("text/html;charset = utf - 8");
        //取得输出对象
        PrintWriter out = response.getWriter();
        request.setCharacterEncoding("GBK");
        //获得part对象
        Part part = 【代码3】
```

```java
        String filediscription = request.getParameter("filediscription");
        out.println("输入的文件描述:" + filediscription + "<br>");
        //指定上传的文件保存到服务器的 uploadFiles 目录中
        File uploadFileDir = new File(getServletContext().getRealPath("/uploadFiles"));
        if(!uploadFileDir.exists()){
            uploadFileDir.mkdir();
        }
        //获得原始文件名
        String oldName = MyUtil.getFileName(part);
        out.println("上传文件的原始名:" + oldName + "<br>");
        out.println("上传文件的大小:" + part.getSize() + "<br>");
        if(oldName != null){
            //上传到服务器的 uploadFiles 目录中
            part.write(uploadFileDir + File.separator + oldName);
        }
        out.println("文件上传到:" + uploadFileDir + File.separator + oldName + "<br>");
    }
}
```

uploadHttpOneServet.java 的响应结果如图 10.2 所示。

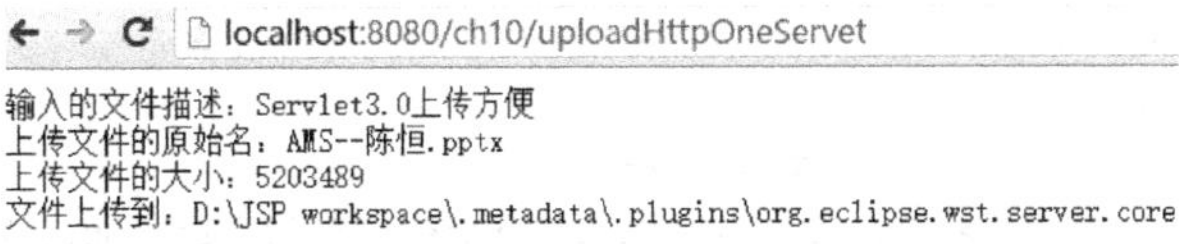

图 10.2 uploadHttpOneServet 的响应结果

3) 任务小结或知识扩展

需要特别注意的是,使用 Eclipse IDE 内嵌的浏览器上传文件时可能会出现 FileNotFoundException 异常。这时使用常规浏览器(如谷歌浏览器)上传文件即可。

4) 代码模板的参考答案

【代码 1】: enctype = "multipart/form-data"
【代码 2】: type = "file"
【代码 3】: request.getPart("resPath");

任务 2:多文件上传

1) 任务的主要内容

编写 uploadHttpMulti.jsp 页面,在页面中选择多个文件提交给 urlPatterns 为{"/uploadHttpMultiServet "}的 servlet 对象(由 UploadHttpMultiServet 类负责创建)。在 Servlet 中进行多文件上传。页面 uploadHttpMulti.jsp 的运行效果如图 10.3 所示。

2) 任务的代码模板

页面文件 uploadHttpMulti.jsp 的代码模板如下:

```jsp
<%@ page language = "java" contentType = "text/html; charset = GBK" pageEncoding = "GBK" %>
<html>
    <head>
        <title>HttpServletRequest 对文件上传的支持</title>
```

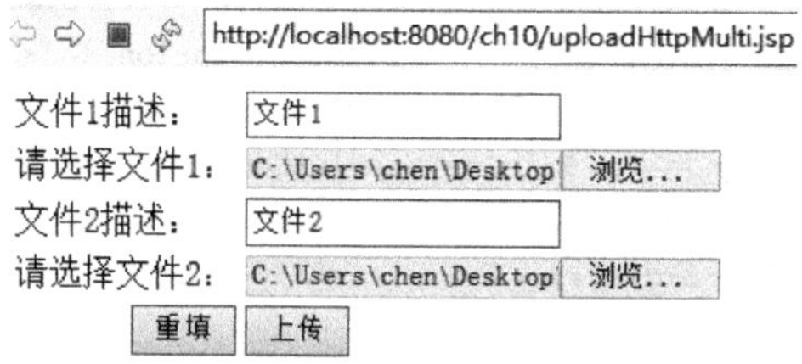

图 10.3　uploadHttpMulti.jsp 页面的运行效果

```
        </head>
        <body>
            <form action = "uploadHttpMultiServet" method = "post" enctype = "multipart/form-data">
                <table>
                    <tr>
                        <td>文件 1 描述：</td>
                        <td><input type = "text" name = "filediscription1"></td>
                    </tr>
                    <tr>
                        <td>请选择文件 1：</td>
                        <td><input type = "file" name = "resPath1"></td>
                    </tr>
                    <tr>
                        <td>文件 2 描述：</td>
                        <td><input type = "text" name = "filediscription2"></td>
                    </tr>
                    <tr>
                        <td>请选择文件 2：</td>
                        <td><input type = "file" name = "resPath2"></td>
                    </tr>
                    <tr>
                        <td align = "right"><input type = "reset" value = "重填"></td>
                        <td><input type = "submit" value = "上传"></td>
                    </tr>
                </table>
            </form>
        </body>
</html>
```

uploadHttpMultiServet.java 的代码模板如下：

```
package servlet;
import java.io.File;
import java.io.IOException;
import java.io.PrintWriter;
import java.util.Collection;
import javax.servlet.ServletException;
import javax.servlet.annotation.MultipartConfig;
import javax.servlet.annotation.WebServlet;
import javax.servlet.http.HttpServlet;
import javax.servlet.http.HttpServletRequest;
import javax.servlet.http.HttpServletResponse;
```

```java
import javax.servlet.http.Part;
import util.MyUtil;
@WebServlet(name = "uploadHttpMultiServet", urlPatterns = { "/uploadHttpMultiServet" })
@MultipartConfig
public class UploadHttpMultiServet extends HttpServlet {
    private static final long serialVersionUID = 1L;
    protected void doGet(HttpServletRequest request, HttpServletResponse response) throws ServletException, IOException {
        doPost(request, response);
    }
    protected void doPost(HttpServletRequest request, HttpServletResponse response) throws ServletException, IOException {
        //设置响应的内容类型
        response.setContentType("text/html;charset=utf-8");
        //取得输出对象
        PrintWriter out = response.getWriter();
        request.setCharacterEncoding("GBK");
        String filediscription1 = request.getParameter("filediscription1");
        out.println("输入的文件 1 描述:" + filediscription1 + "<br>");
        String filediscription2 = request.getParameter("filediscription2");
        out.println("输入的文件 2 描述:" + filediscription2 + "<br>");
        //指定上传的文件保存到服务器的 uploadFiles 目录中
        File uploadFileDir = new File(getServletContext().getRealPath("/uploadFiles"));
        if(!uploadFileDir.exists()){
            uploadFileDir.mkdir();
        }
        //获得 Part 集合
        Collection<Part> parts = 【代码 1】
        for (Part part : parts) {
            //没有选择文件或不是文件域
            if (part == null || !part.getName().contains("resPat")) {
                continue;
            }
            //获得原始文件名
            String oldName = MyUtil.getFileName(part);
            out.println("上传文件的原始名:" + oldName + "<br>");
            out.println("上传文件的大小:" + part.getSize() + "<br>");
            if(oldName != null){
                //上传到服务器的 uploadFiles 目录中
                part.write(uploadFileDir + File.separator + oldName);
            }
            out.println("文件上传到:" + uploadFileDir + File.separator + oldName + "<br>");
        }
    }
}
```

uploadHttpMultiServet.java 的响应结果如图 10.4 所示。

3）任务小结或知识扩展

上传多个文件实际上只是迭代上传单个文件而已，与单个文件上传没有本质区别。

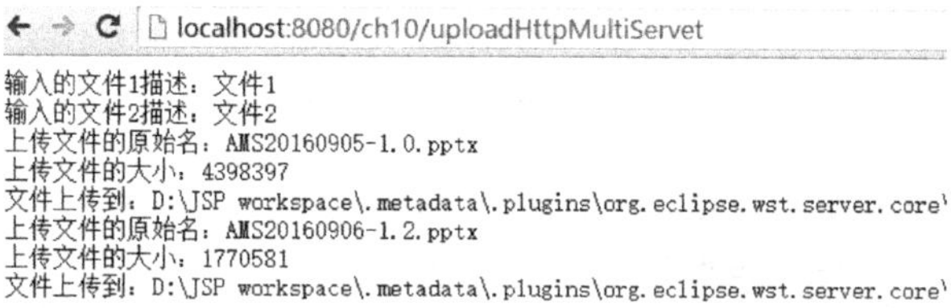

图 10.4　uploadHttpMultiServet 的响应结果

4）代码模板的参考答案

【代码 1】：request.getParts();

10.1.4　实践环节

尝试使用 jspsmart 或 commons-fileupload 组件进行文件上传。

10.2　文件的下载

10.2.1　核心知识

实现文件下载经常有两种方法：一是通过超链接实现下载；二是利用程序编码实现下载。超链接实现下载固然简单，但暴露了下载文件的真实位置，并且只能下载存放在 Web 应用程序所在的目录下的文件。利用程序编码实现下载，可以增加安全访问控制，还可以从任意位置提供下载的数据，可以将文件存放到 Web 应用程序以外的目录中，也可以将文件保存到数据库中。

利用程序实现下载需要设置以下两个报头。

（1）Web 服务器需要告诉浏览器其所输出内容的类型不是普通文本文件或 HTML 文件，而是一个要保存到本地的下载文件。设置 Content-Type 的值为 application/x-msdownload。

（2）Web 服务器希望浏览器不直接处理相应的实体内容，而是由用户选择将相应的实体内容保存到一个文件中，这需要设置 Content-Disposition 报头。该报头指定了接收程序处理数据内容的方式，在 HTTP 应用中只有 attachment 是标准方式，attachment 表示要求用户干预。在 attachment 后面还可以指定 filename 参数，该参数是服务器建议浏览器将实体内容保存到文件中的文件名称。

设置报头的示例如下：

```
response.setHeader("Content-Type", "application/x-msdownload" );
response.setHeader("Content-Disposition", "attachment;filename=" + filename);
```

10.2.2　能力目标

掌握利用程序实现下载的流程。

10.2.3 任务驱动

下面通过一个实例讲述利用程序实现下载的过程。

该实例要求文件上传到 D:\uploadFile 目录后,把文件名保存到数据库中,然后根据数据库中的文件名从 D:\uploadFile 目录中下载文件。首先运行 uploadCourse.jsp 页面,具体开发步骤如下。

(1) 创建数据表 coursetable(数据库软件为 Oracle 10g)。

```
drop table coursetable;
create table coursetable (
        courseID varchar(17) not null,
        courseName varchar(100) not null,
        coursedis varchar(500) not null,
        resPath varchar(200) not null,
        constraint pk_coursetable primary key (courseid)
);
commit;
```

(2) 编写选择文件页面——uploadCourse.jsp。

页面 uploadCourse.jsp 运行效果如图 10.5 所示。

图 10.5 页面 uploadCourse.jsp 运行效果

页面文件 uploadCourse.jsp 的代码如下:

```
<%@ page language = "java" contentType = "text/html; charset = GBK" pageEncoding = "GBK" %>
<html>
    <head>
        <title>uploadCourse.jsp</title>
    </head>
    <body>
        <form action = "uploadCourse" method = "post" enctype = "multipart/form-data">
            <table>
                <tr>
                    <td>课程名称:</td>
                    <td><input type = "text" name = "courseName"></td>
                </tr>
```

```html
                <tr>
                    <td>课程描述：</td>
                    <td>
                        <textarea rows="10" cols="30" name="coursedis"></textarea>
                    </td>
                </tr>
                <tr>
                    <td>请选择课件：</td>
                    <td><input type="file" name="resPath"></td>
                </tr>
                <tr>
                    <td align="right"><input type="reset" value="重填"></td>
                    <td><input type="submit" value="上传"></td>
                </tr>
            </table>
        </form>
    </body>
</html>
```

（3）编写上传文件 Servlet——uploadCourseServlet.java。uploadCourseServlet.java 的代码如下：

```java
package servlet;
import java.io.File;
import java.io.IOException;
import java.sql.Connection;
import java.sql.DriverManager;
import java.sql.PreparedStatement;
import javax.servlet.RequestDispatcher;
import javax.servlet.ServletException;
import javax.servlet.annotation.MultipartConfig;
import javax.servlet.annotation.WebServlet;
import javax.servlet.http.HttpServlet;
import javax.servlet.http.HttpServletRequest;
import javax.servlet.http.HttpServletResponse;
import javax.servlet.http.Part;
import util.MyUtil;
@WebServlet(name = "uploadCourse", urlPatterns = { "/uploadCourse" })
@MultipartConfig(maxFileSize = 10 * 1024 * 1024)//设置上传文件的最大值为10MB
public class UploadCourseServlet extends HttpServlet {
    private static final long serialVersionUID = 1L;
    protected void doGet(HttpServletRequest request, HttpServletResponse response) throws ServletException, IOException {
        doPost(request, response);
    }
    protected void doPost(HttpServletRequest request, HttpServletResponse response) throws ServletException, IOException {
        Connection con = null;
        PreparedStatement pst = null;
        //设置响应的内容类型
```

```java
response.setContentType("text/html;charset=utf-8");
request.setCharacterEncoding("GBK");
//成功后,显示出课程信息
RequestDispatcher dis1 = request.getRequestDispatcher("showInfo");
//失败页面
RequestDispatcher dis2 = request.getRequestDispatcher("uploadFail.jsp");
//重新回到上传页面
RequestDispatcher dis3 = request.getRequestDispatcher("uploadCourse.jsp");
try {
    //指定上传的文件保存到D盘的uploadFile目录中
    File uploadFileDir = new File("D:\\uploadFile");
    if(!uploadFileDir.exists()){
        uploadFileDir.mkdir();
    }
    //获得part对象
    Part part = request.getPart("resPath");
    //获得原始文件名
    String oldName = MyUtil.getFileName(part);
    //如果没有选择课件,回到上传页面继续上传
    if((null == oldName || "".equals(oldName)) && part.getSize() == 0){
        dis3.forward(request, response);
    }
    //如果传送的文件名是全路径名,取出原始文件名
    int index = oldName.lastIndexOf(File.separator);
    if(index > 0){
        oldName = oldName.substring(index + 1, oldName.length());
    }
    //获取文件类型
    String fileType = oldName.substring(oldName.lastIndexOf("."));
    //以一定规范的文件名上传,因为客户上传时,文件名有可能重名
    String newName = MyUtil.getStringID() + fileType;
    //上传到服务器的uploadFile目录中
    part.write(uploadFileDir + File.separator + oldName);
    //把文件信息写入数据库
    Class.forName("oracle.jdbc.driver.OracleDriver");
    //获得数据库连接
    con = DriverManager.getConnection("jdbc:oracle:thin:@localhost:1521:orcl","system","system");
    String sql = "insert into coursetable values(?,?,?,?)";
    pst = con.prepareStatement(sql);
    pst.setString(1, MyUtil.getStringID());
    pst.setString(2, request.getParameter("courseName"));
    pst.setString(3, request.getParameter("coursedis"));
    pst.setString(4, newName);
    pst.executeUpdate();
    pst.close();
    con.close();
    dis1.forward(request, response);
} catch (Exception e) {
    //TODO Auto-generated catch block
    e.printStackTrace();
```

```
            dis2.forward(request, response);
        }
    }
}
```

(4) 编写查询文件信息 Servlet——ShowInfoServlet.java。

ShowInfoServlet.java 的代码如下：

```java
package servlet;
import java.io.IOException;
import java.sql.Connection;
import java.sql.DriverManager;
import java.sql.PreparedStatement;
import java.sql.ResultSet;
import java.util.ArrayList;
import javax.servlet.RequestDispatcher;
import javax.servlet.ServletException;
import javax.servlet.annotation.WebServlet;
import javax.servlet.http.HttpServlet;
import javax.servlet.http.HttpServletRequest;
import javax.servlet.http.HttpServletResponse;
import dto.Course;
@WebServlet(name = "showInfo", urlPatterns = { "/showInfo" })
public class ShowInfoServlet extends HttpServlet {
    private static final long serialVersionUID = 1L;
    protected void doGet(HttpServletRequest request, HttpServletResponse response) throws
    ServletException, IOException {
        doPost(request, response);
    }
    protected void doPost(HttpServletRequest request, HttpServletResponse response)
    throws ServletException, IOException {
        Connection con = null;
        PreparedStatement pst = null;
        ResultSet rs = null;
        //存放课程的集合 ArrayList
        ArrayList<Course> al = new ArrayList<Course>();
        try {
            Class.forName("oracle.jdbc.driver.OracleDriver");
            //获得数据库连接
            con = DriverManager.getConnection("jdbc:oracle:thin:@localhost:1521:
            orcl","system","system");
            String sql = "select * from coursetable";
            pst = con.prepareStatement(sql);
            rs = pst.executeQuery();
            //遍历出结果集中的课程
            while(rs.next()){
                Course c = new Course();
                c.setCourseID(rs.getString("courseID"));
                c.setCourseName(rs.getString("courseName"));
                c.setCoursedis(rs.getString("coursedis"));
                c.setResPath(rs.getString("resPath"));
```

```
            al.add(c);
        }
        rs.close();
        pst.close();
        con.close();
        //存储在 request 中转发到 showInfo.jsp
        request.setAttribute("courses", al);
        RequestDispatcher dis1 = request.getRequestDispatcher("showInfo.jsp");
        dis1.forward(request, response);
    } catch (Exception e) {
        //TODO Auto-generated catch block
        e.printStackTrace();
    }
  }
}
```

(5) 编写文件显示页面——showInfo.jsp。

页面 showInfo.jsp 的运行效果如图 10.6 所示。

图 10.6　页面 showInfo.jsp 运行效果

页面文件 showInfo.jsp 的代码如下：

```jsp
<%@ page language="java" contentType="text/html; charset=GBK" pageEncoding="GBK"%>
<%@ taglib prefix="c" uri="http://java.sun.com/jsp/jstl/core"%>
<html>
    <head>
        <title>showInfo.jsp</title>
    </head>
    <body>
        <table border="1">
            <tr bgcolor="LightGreen">
                <th>课程 ID</th>
                <th>课程名称</th>
                <th>课程描述</th>
                <th>下载课件</th>
            </tr>
            <c:forEach var="course" items="${requestScope.courses}">
                <tr>
                    <td>${course.courseID}</td>
                    <td>${course.courseName}</td>
                    <td>${course.coursedis}</td>
                    <td align="center"><a href="download?resPath=${course.resPath}" style=
                    "text-decoration:none">课件</a></td>
                </tr>
            </c:forEach>
```

```
        </table>
    </body>
</html>
```

(6) 编写下载 Servlet——DownloadServlet.java。

DownloadServlet.java 的代码如下：

```java
package servlet;
import java.io.FileInputStream;
import java.io.IOException;
import javax.servlet.ServletException;
import javax.servlet.ServletOutputStream;
import javax.servlet.annotation.WebServlet;
import javax.servlet.http.HttpServlet;
import javax.servlet.http.HttpServletRequest;
import javax.servlet.http.HttpServletResponse;
@WebServlet(name = "download", urlPatterns = { "/download" })
public class DownloadServlet extends HttpServlet {
    private static final long serialVersionUID = 1L;
    protected void doGet(HttpServletRequest request, HttpServletResponse response) throws
    ServletException, IOException {
        doPost(request, response);
    }
    protected void doPost(HttpServletRequest request, HttpServletResponse response)
    throws ServletException, IOException {
        String aFilePath = null;            //要下载的文件路径
        String aFileName = null;            //要下载的文件名
        FileInputStream in = null;          //输入流
        ServletOutputStream out = null;     //输出流
        try {
            aFilePath = "D:\\uploadFile\\";
            aFileName = request.getParameter("resPath");
            //设置下载文件使用的报头
            response.setHeader("Content - Type", "application/x - msdownload" );
            response.setHeader("Content - Disposition", "attachment; filename = "
                + aFileName);
            //读入文件
            in = new FileInputStream(aFilePath + aFileName);
            //得到响应对象的输出流,用于向客户端输出二进制数据
            out = response.getOutputStream();
            out.flush();
            int aRead = 0;
            byte b[] = new byte[1024];
            while ((aRead = in.read(b)) != -1 & in != null) {
                out.write(b,0,aRead);
            }
            out.flush();
            in.close();
            out.close();
        } catch (Throwable e) {
            e.printStackTrace();
```

        }
    }
}

如果下载的文件名中有中文字符,浏览器提示保存的文件名将显示为乱码。要解决这个乱码问题,需要对下载的文件名按照 UTF-8 进行编码。在本章工具类 MyUtil 中添加一个静态的字符编码转换方法 String toUTF8String(String str)。

修改本节的 DownloadServlet.java,对要下载的文件名调用 toUTF8String() 方法,代码片段如下:

```
…
aFileName = request.getParameter("resPath");
//设置下载文件使用的报头
response.setHeader("Content-Type", "application/x-msdownload" );
response.setHeader("Content-Disposition", "attachment; filename = " + MyUtil.toUTF8String(aFileName));
//读入文件
…
```

10.2.4 实践环节

将 10.2.3 节中数据表 coursetable 的某条记录的 resPath 列对应文件名修改为中文,然后到 D:\\uploadFile 目录下,把某个文件名修改为相同的中文名。运行程序,下载该文件。查看浏览器提示保存文件名是否为乱码,如果是,如何解决乱码问题?

10.3 小　　结

本章主要介绍了基于 Servlet 3.0 上传文件的使用方法,列举了文件上传与下载的实例。

习　题　10

1. 基于表单的文件上传,需要注意什么?
2. 文件下载的实现方法有哪些? 它们的优缺点是什么?

地址簿管理信息系统

主要内容

(1) 基于 Servlet MVC 模式开发 Web 应用。
(2) 小型管理信息系统的开发技巧。

本章通过一个小型的地址簿管理信息系统,讲述如何采用基于 Servlet MVC 模式来开发一个 Web 应用。系统的开发环境如下。

(1) 操作系统:Windows 10。
(2) 数据库:Oracle 10g。
(3) JDK:1.7。
(4) JSP 引擎:Tomcat 7.0。
(5) 集成开发环境(IDE):Eclipse IDE for Java EE Developers。

11.1 系 统 设 计

11.1.1 系统功能需求

地址簿管理信息系统是提供给注册用户使用的系统。系统提供的功能如下。
(1) 非注册用户可以注册为注册用户。
(2) 成功注册的用户,可以登录系统。
(3) 成功登录的用户,可以添加、修改、删除以及浏览自己的朋友信息。
(4) 成功登录的用户,可以修改自己的登录密码。

11.1.2 系统模块划分

注册用户使用地址簿管理信息系统可以添加、修改、删除以及查询自己的朋友信息,具体系统功能模块如下。

1. 用户注册

新用户填写注册信息,包括用户名、密码和确认密码。输入用户名时,系统会提示用户名是否可用。

2. 用户登录

用户输入用户名、密码进行登录。登录失败，系统回到登录画面继续登录。登录成功，进入系统管理主页面(main.jsp)，包括添加朋友信息，修改朋友信息，删除朋友信息，查询朋友信息，修改密码以及退出系统等功能。

3. 添加朋友信息

用户填写朋友信息表单，包括朋友姓名、生日、电话、E-mail、照片、地址以及关系等信息。提交朋友信息表单时，系统使用JavaScript验证信息是否输入以及信息格式是否合法。

4. 修改朋友信息

系统首先根据成功登录的用户名查询出该用户的所有朋友信息，然后用户选择某个朋友进行修改信息。

5. 删除朋友信息

系统首先根据成功登录的用户名查询出该用户的所有朋友信息，然后用户选择某个或多个朋友进行删除。

6. 查询朋友信息

系统根据成功登录的用户名查询出该用户的所有朋友信息。

7. 修改密码

成功登录的用户，从主页面进入该页面修改自己的密码。

8. 退出系统

成功登录的用户，在主页面单击"退出系统"链接，系统首先清除用户的会话(Session)，然后返回登录页面。

11.2 数据库设计

系统采用加载纯Java数据库驱动程序的方式连接Oracle 10g。在Oracle 10g的默认数据库orcl中创建两张表：usertable与friendinfo。

11.2.1 数据库概念结构设计

根据系统设计与分析，可以设计出如下的数据结构。

1. 用户信息

包括用户名和密码等，一个用户可以添加多个朋友信息。

2. 朋友信息

包括朋友ID、姓名、生日、电话、E-mail、照片、地址、关系以及所属的用户等信息。

根据以上的数据结构，结合数据库设计的特点，可以画出如图11.1所示的数据库概念结构图。

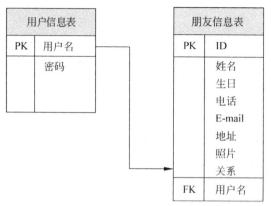

图 11.1 数据库概念结构图

11.2.2 数据库逻辑结构设计

将数据库概念结构图转换为 Oracle 数据库所支持的实际数据模型，即数据库的逻辑结构。系统中两张表的设计如表 11.1 和表 11.2 所示。

表 11.1 用户信息表（usertable）

字段	含义	类型	长度	是否为空
userName	用户名	varchar	20	no
password	密码	varchar	20	no

表 11.2 朋友信息表（friendinfo）

字段	含义	类型	长度	是否为空
id	朋友编号	varchar	17	no
name	朋友姓名	varchar	20	no
birthday	朋友生日	date	—	yes
telephone	朋友电话	varchar	20	yes
email	朋友 E-mail	varchar	20	yes
address	朋友地址	varchar	50	yes
picture	朋友照片	varchar	25	yes
relation	关系	varchar	10	no
userName	用户名	varchar	20	no

11.2.3 创建数据表

根据数据库的逻辑结构，创建数据表的代码如下：

```
drop table friendinfo;
drop table usertable;
create table usertable (
    userName varchar(20) not null,
    password varchar(20) not null,
    constraint pk_usertable primary key (userName)
```

```sql
);
create table friendinfo (
    id varchar(17) not null,
    name varchar(20) not null,
    birthday date null,
    telephone varchar(20) null,
    email varchar(20) null,
    address varchar(50) null,
    picture varchar(25) null,
    relation varchar(10) not null,
    userName varchar(20) not null,
    constraint pk_friendinfo primary key (id),
        constraint fk_friendinfo_1 foreign key (userName)
        references usertable (userName)
);
commit;
```

11.3 系统管理

11.3.1 导入相关的jar包

在本章中新建一个 Web 工程 ch11，在 ch11 工程中开发本系统。由于在本系统所有 JSP 页面中尽量使用 EL 表达式和 JSTL 标签，又因为本系统采用纯 Java 数据库驱动程序连接 Oracle 10g。所以，需要将 standard.jar、jstl.jar 和 classes12.jar 复制到 ch11/WEB-INF/lib 文件夹中。具体操作步骤请参考本书 6.2 节和 9.2 节。

11.3.2 JSP页面管理

本系统所有 JSP 页面（包括相关的 JavaScript 文件）都保存在 ch11 的 WebContent 目录中。有些 JSP 页面使用到 CSS 与 JavaScript。有关 CSS 与 JavaScript 的知识超出了本书的讨论范围，请读者查阅相关书籍。

本系统中除了登录、注册以及修改密码等操作不在系统管理主页面（main.jsp）中实现，其他操作都在 main.jsp 中实现。在该页面中使用 DIV＋CSS＋Iframe 进行布局管理。

用户首先通过 http://localhost:8080/ch11/login.jsp 访问登录页面，登录成功，进入系统管理主页面，main.jsp 运行效果如图 11.2 所示。

main.jsp 的核心代码如下：

```html
<body>
<div id="top">
    <br>
    <br>
    <center>
        <font size=6 face=华文新魏>欢迎 ${sessionScope.user.uname}使用地址簿管理系统
        </font>
    </center>
</div>
<div id="left">
```

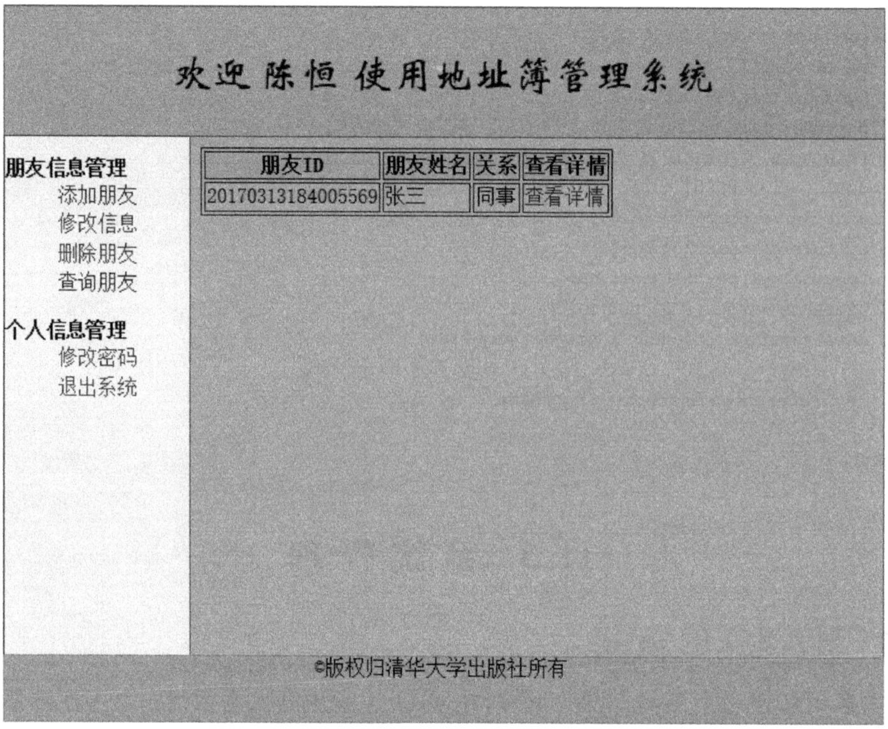

图 11.2 系统管理主页面

```
<dl>
    <dt onClick = 'showHide("items1_1")'>
        <b>朋友信息管理</b>
    </dt>
    <dd style = 'display: block' id = 'items1_1'>
        <table>
            <tr>
                <td><a href = "addFriend.jsp" target = "right"
                    style = "text-decoration: none">添加朋友</a>
                </td>
            </tr>
            <tr>
                <td><a href = "queryFriendServlet?flag = update" target = "right"
                    style = "text-decoration: none">修改信息</a>
                </td>
            </tr>
            <tr>
                <td><a href = "queryFriendServlet?flag = del" target = "right"
                    style = "text-decoration: none">删除朋友</a>
                </td>
            </tr>
            <tr>
                <td><a href = "queryFriendServlet" target = "right"
                    style = "text-decoration: none">查询朋友</a>
                </td>
```

```
              </tr>
            </table>
          </dd>
        </dl>
        <dl>
          <dt onClick='showHide("items1_2")'>
            <b>个人信息管理</b>
          </dt>
          <dd style='display:block' id='items1_2'>
            <table>
              <tr>
                <td><a href="upadatepassword.jsp"
                    style="text-decoration:none">修改密码</a>
                </td>
              </tr>

              <tr>
                <td><a href="exitUserServlet"
                    style="text-decoration:none">退出系统</a>
                </td>
              </tr>
            </table>
          </dd>
        </dl>
      </div>
      <div id="right1">
        <iframe src="queryFriendServlet" name="right" width="100%" height="100%"
          frameborder="0"></iframe>
      </div>
      <div id="bottom">
        <center>&copy;版权归清华大学出版社所有</center>
      </div>
    </body>
```

11.3.3 组件与 Servlet 管理

本系统使用的组件与 servlet 包层次结构如图 11.3 所示。

1. busyness 包

图 11.3 中 busyness 包里存放的 Java 程序都是实现业务逻辑处理的 bean(业务模型)，包括用户注册、登录、朋友信息增、删、改、查等业务处理。

2. common 包

该包中 MyUtil 类的 getStringID()方法是主键产生器，DBConnection 类用于实现数据库连接与关闭等操作。

3. entity 包

该包的 Friend 类和 User 类是实现数据封装的实体 bean(实体模型)。

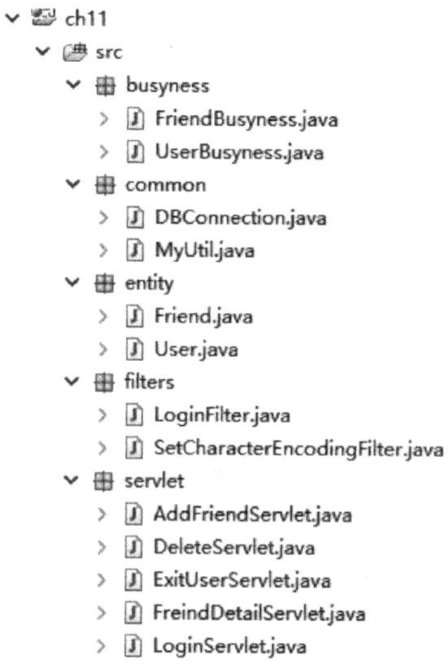

```
∨ ⊞ ch11
  ∨ ⊞ src
    ∨ ⊞ busyness
      > ⓙ FriendBusyness.java
      > ⓙ UserBusyness.java
    ∨ ⊞ common
      > ⓙ DBConnection.java
      > ⓙ MyUtil.java
    ∨ ⊞ entity
      > ⓙ Friend.java
      > ⓙ User.java
    ∨ ⊞ filters
      > ⓙ LoginFilter.java
      > ⓙ SetCharacterEncodingFilter.java
    ∨ ⊞ servlet
      > ⓙ AddFriendServlet.java
      > ⓙ DeleteServlet.java
      > ⓙ ExitUserServlet.java
      > ⓙ FreindDetailServlet.java
      > ⓙ LoginServlet.java
```

图 11.3 包层次结构图

4. filters 包

系统中使用过滤器(filter)解决中文乱码和登录验证等问题,系统中过滤器 Java 代码的实现都存放在 filters 包中。

5. servlet 包

系统中控制器的实现都存放在 servlet 包中。

11.4 组件设计

系统使用到的组件有过滤器、数据库操作、实体模型(数据封装 bean)和业务模型(业务处理 bean)4 部分。

11.4.1 过滤器

系统使用了两个过滤器:设置字符编码过滤器与登录验证过滤器。

1. 设置字符编码过滤器

系统使用过滤器解决中文乱码问题。当用户提交请求时,在请求处理之前,系统使用过滤器把用户提交的信息进行解码与编码,避免了乱码出现。具体代码如下。

SetCharacterEncodingFilter.java 的核心代码如下:

```java
//过滤器设置字符编码为 GBK
@WebFilter(filterName = "setCharacterEncodingFilter", urlPatterns = { "/*" })
public class SetCharacterEncodingFilter implements Filter {
    public void destroy() {
    }
    public void doFilter(ServletRequest request, ServletResponse response,
            FilterChain chain) throws IOException, ServletException {
        request.setCharacterEncoding("GBK");
```

```java
        //执行下一个过滤器
        chain.doFilter(request, response);
    }
    public void init(FilterConfig filterConfig) throws ServletException {

    }
}
```

2. 登录验证过滤器

从系统分析得知,只有用户成功登录了,才能使用该系统。也就是说,用户不能成功登录时,不允许使用除登录、注册之外的功能。当用户通过 URL 请求时,系统首先使用过滤器判别用户访问的是否是登录或注册的功能,如果不是,就判断用户的会话(session)是否存在,如果不存在,就提示用户先登录。具体代码如下。

LoginFilter.java 的核心代码如下:

```java
@WebFilter(filterName = "loginFilter", urlPatterns = { "/*" })
public class LoginFilter implements Filter {
    public void destroy() {
    }
    public void doFilter(ServletRequest request, ServletResponse response,
            FilterChain chain) throws IOException, ServletException {
        HttpServletRequest req = (HttpServletRequest) request;
        HttpServletResponse resp = (HttpServletResponse) response;
        resp.setContentType("text/html;");
        resp.setCharacterEncoding("GBK");
        HttpSession session = req.getSession();
        //得到用户请求的 URI
        String request_uri = req.getRequestURI();
        //得到 Web 应用程序的上下文路径
        String ctxPath = req.getContextPath();
        //去除上下文路径,得到剩余部分的路径
        String uri = request_uri.substring(ctxPath.length());
        //判断用户访问的是否是登录页面或注册页面
        if (uri.equals("/login.jsp") || uri.equals("/regist.jsp")
                || uri.equals("/loginServlet") || uri.equals("/registServlet")) {
            //执行下一个过滤器
            chain.doFilter(request, response);
        } else {
            //如果访问的不是登录页面,则判断用户是否已经登录
            if (null != session.getAttribute("user")
                    && "" != session.getAttribute("user")) {
                chain.doFilter(request, response);
            } else {
                PrintWriter out = resp.getWriter();
                String pah = ctxPath + "/login.jsp";
                out.println("您没有登录,请先<a href=" + pah + " target=_top>登录</a>!");
                return;
            }
        }
```

```
        }
    }
    public void init(FilterConfig config) throws ServletException {
    }
}
```

11.4.2 数据库操作

1. 主键生成器

朋友信息表的主键(ID)产生策略是,格式化系统时间,主键格式(共 17 位)如下:

yyyyMMddHHmmssSSS

主键生成器代码如下:

```
public static String getStringID(){
    String id = null;
    Date date = new Date();
    SimpleDateFormat sdf = new SimpleDateFormat("yyyyMMddHHmmssSSS");
    id = sdf.format(date);
    return id;
}
```

2. 数据库连接与关闭

数据库连接与关闭是由 common 包中的 DBConnection 类实现的。类中方法功能说明如下。

(1) public synchronized static Connection getOneCon():从连接池中获得一个连接对象。

(2) public static void close(ResultSet rs):关闭结果集对象。

(3) public static void close(PreparedStatement ps):关闭预处理对象。

(4) public synchronized static void close(Connection con):把连接对象放回连接池中。

DBConnection.java 的代码如下:

```
package common;
import java.sql.*;
import java.util.ArrayList;
public class DBConnection {
    //存放 Connection 对象的数组,数组被看成连接池
    static ArrayList<Connection> list = new ArrayList<Connection>();
    //从连接池中取出一个连接对象
    public synchronized static Connection getOneCon(){
        //如果连接池中有连接对象
        if(list.size()>0){
            return list.remove(0);
        }
        //连接池没有连接对象则创建连接放到连接池中
        else{
            for(int i = 0; i<5; i++){
```

```
                try {
                    Class.forName("oracle.jdbc.driver.OracleDriver");
                    Connection con = DriverManager.
                        getConnection("jdbc:oracle:thin:@localhost:1521:orcl","system",
                        "system");
                    list.add(con);
                } catch (Exception e) {
                    //TODO Auto-generated catch block
                    e.printStackTrace();
                }
            }
            return list.remove(0);
        }
    }
    //关闭结果集对象
    public static void close(ResultSet rs){
        try {
            if(rs != null)
                rs.close();
        } catch (SQLException e) {
            //TODO Auto-generated catch block
            e.printStackTrace();
        }
    }
    //关闭预处理语句
    public static void close(PreparedStatement ps){
        try {
            if(ps! = null)
                ps.close();
        } catch (SQLException e) {
            //TODO Auto-generated catch block
            e.printStackTrace();
        }
    }
    //把连接对象放回连接池中
    public synchronized static void close(Connection con){
        if(con != null)
            list.add(con);
    }
}
```

11.4.3 实体模型

实体模型主要用于封装 JSP 页面提交的信息以及与数据库交互的数据传递。系统共使用两个实体模型：用户(User)和朋友(Friend)。

11.4.4 业务模型

系统共用到两个业务 bean：FriendBusyness 和 UserBusyness。和朋友信息有关的业务逻辑处理都写在 FriendBusyness 类里,和用户有关的业务逻辑处理都写在 UserBusyness

类里。

1. FriendBusyness 类

FriendBusyness 类的各方法功能说明如下。

(1) public boolean addFriend(Friend f,String userName)：添加朋友信息。

(2) public ArrayList<Friend> getAllFriends(String userName)：根据用户名查询该用户的朋友信息。

(3) public Friend getAFriend(String id)：根据朋友的 ID 查询该朋友的信息。

(4) public void deleteFriend(String id[])：根据朋友的 ID 删除朋友信息，删除一个或多个朋友。

(5) public boolean upadateFriend(Friend f)：更新朋友信息。

FriendBusyness.java 的代码如下：

```java
package busyness;
import java.sql.Connection;
import java.sql.PreparedStatement;
import java.sql.ResultSet;
import java.sql.SQLException;
import java.util.ArrayList;
import common.DBConnection;
import entity.Friend;
public class FriendBusyness {
    //实现添加朋友信息
    public boolean addFriend(Friend f, String userName) {
        boolean b = false;
        Connection con = DBConnection.getOneCon();
        PreparedStatement ps = null;
        try {
            ps = con.prepareStatement("insert into friendinfo
            values(?,?,to_date(?,'YYYY-MM-DD'),?,?,?,?,?,?)");
            ps.setString(1, f.getId());
            ps.setString(2, f.getName());
            ps.setString(3, f.getBirthday());   //日期类型
            ps.setString(4, f.getTelephone());
            ps.setString(5, f.getEmail());
            ps.setString(6, f.getAddress());
            ps.setString(7, f.getPicture());
            ps.setString(8, f.getRelation());
            ps.setString(9, userName);
            int i = ps.executeUpdate();
            if (i > 0)
                b = true;
        } catch (SQLException e) {
            //TODO Auto-generated catch block
            e.printStackTrace();
        } finally {
            DBConnection.close(ps);
            DBConnection.close(con);
```

```java
        }
        return b;
    }
    //查询朋友信息
    public ArrayList<Friend> getAllFriends(String userName) {
        ArrayList<Friend> al = new ArrayList<Friend>();
        Connection con = DBConnection.getOneCon();
        PreparedStatement ps = null;
        ResultSet rs = null;
        try {
            ps = con.prepareStatement("select * from friendinfo where userName = ?");
            ps.setString(1, userName);
            rs = ps.executeQuery();
            while (rs.next()) {
                Friend f = new Friend();
                f.setId(rs.getString(1));
                f.setName(rs.getString(2));
                f.setBirthday(rs.getString(3).substring(0, 10));      //取出 YYYY-MM-DD
                f.setTelephone(rs.getString(4));
                f.setEmail(rs.getString(5));
                f.setAddress(rs.getString(6));
                f.setPicture(rs.getString(7));
                f.setRelation(rs.getString(8));
                al.add(f);
            }
        } catch (SQLException e) {
            //TODO Auto-generated catch block
            e.printStackTrace();
        } finally {
            DBConnection.close(rs);
            DBConnection.close(ps);
            DBConnection.close(con);
        }
        return al;
    }
    //查询一个朋友信息
    public Friend getAFriend(String id) {
        Connection con = DBConnection.getOneCon();
        PreparedStatement ps = null;
        ResultSet rs = null;
        Friend f = new Friend();
        try {
            ps = con.prepareStatement("select * from friendinfo where id = ?");
            ps.setString(1, id);
            rs = ps.executeQuery();
            if (rs.next()) {
                f.setId(rs.getString(1));
                f.setName(rs.getString(2));
                f.setBirthday(rs.getString(3).substring(0, 10));      //取出 YYYY-MM-DD
                f.setTelephone(rs.getString(4));
                f.setEmail(rs.getString(5));
```

```java
                    f.setAddress(rs.getString(6));
                    f.setPicture(rs.getString(7));
                    f.setRelation(rs.getString(8));
                }
            } catch (SQLException e) {
                //TODO Auto-generated catch block
                e.printStackTrace();
            } finally {
                DBConnection.close(rs);
                DBConnection.close(ps);
                DBConnection.close(con);
            }
            return f;
        }
        //删除朋友信息
        public void deleteFriend(String id[]) {
            Connection con = DBConnection.getOneCon();
            PreparedStatement ps = null;
            try {
                ps = con.prepareStatement("delete from friendinfo where id = ?");
                for (int i = 0; i < id.length; i++) {
                    ps.setString(1, id[i]);
                    ps.executeUpdate();
                }
            } catch (SQLException e) {
                //TODO Auto-generated catch block
                e.printStackTrace();
            } finally {
                DBConnection.close(ps);
                DBConnection.close(con);
            }
        }
        //修改朋友信息
        public boolean upadateFriend(Friend f) {
            boolean b = false;
            Connection con = DBConnection.getOneCon();
            PreparedStatement ps = null;
            try {
                String sql = "update friendinfo set name = ?,"
                        + "birthday = to_date(?,'YYYY-MM-DD'),"
                        + "telephone = ?,"
                        + "email = ?,"
                        + "address = ?,";
                if (f.getPicture() != null) {//修改了照片
                    sql = sql + "picture = ?,";
                }
                sql = sql + "relation = ? where id = ?";
                ps = con.prepareStatement(sql);
                ps.setString(1, f.getName());
                ps.setString(2, f.getBirthday());           //日期类型
                ps.setString(3, f.getTelephone());
```

```java
            ps.setString(4, f.getEmail());
            ps.setString(5, f.getAddress());
            if (f.getPicture() != null) {
                ps.setString(6, f.getPicture());
                ps.setString(7, f.getRelation());
                ps.setString(8, f.getId());
            } else {
                ps.setString(6, f.getRelation());
                ps.setString(7, f.getId());
            }
            int i = ps.executeUpdate();
            if (i > 0)
                b = true;
        } catch (SQLException e) {
            //TODO Auto-generated catch block
            e.printStackTrace();
        } finally {
            DBConnection.close(ps);
            DBConnection.close(con);
        }
        return b;
    }
}
```

2. UserBusyness 类

UserBusyness 类的各方法功能说明如下。

(1) public boolean nameIsExit(String name)：判断用户名是否存在。

(2) public boolean regist(User u)：注册用户信息。

(3) public boolean login(User u)：实现登录功能。

(4) public boolean upadatePassword(User u)：更新用户的密码。

UserBusyness.java 的代码如下：

```java
package busyness;
import java.sql.*;
import common.DBConnection;
import entity.User;
public class UserBusyness {
    //判断用户名是否可用
    public boolean nameIsExit(String name) {
        boolean b = true;
        Connection con = DBConnection.getOneCon();
        PreparedStatement ps = null;
        ResultSet rs = null;
        try {
            ps = con.prepareStatement("select * from usertable where userName=?");
            ps.setString(1, name);
            rs = ps.executeQuery();
            if (rs.next())
                b = false;
```

```java
        } catch (SQLException e) {
            //TODO Auto-generated catch block
            e.printStackTrace();
        } finally {
            DBConnection.close(rs);
            DBConnection.close(ps);
            DBConnection.close(con);
        }
        return b;
    }
    //实现注册功能
    public boolean regist(User u) {
        boolean b = false;
        Connection con = DBConnection.getOneCon();
        PreparedStatement ps = null;
        try {
            ps = con.prepareStatement("insert into usertable values(?,?)");
            ps.setString(1, u.getUname());
            ps.setString(2, u.getUpass());
            int i = ps.executeUpdate();
            if (i > 0)
                b = true;
        } catch (SQLException e) {
            //TODO Auto-generated catch block
            e.printStackTrace();
        } finally {
            DBConnection.close(ps);
            DBConnection.close(con);
        }
        return b;
    }
    //实现登录功能
    public boolean login(User u) {
        boolean b = false;
        Connection con = DBConnection.getOneCon();
        PreparedStatement ps = null;
        ResultSet rs = null;
        try {
            ps = con.prepareStatement("select * from usertable where userName = ? and password = ?");
            ps.setString(1, u.getUname());
            ps.setString(2, u.getUpass());
            rs = ps.executeQuery();
            if (rs.next())
                b = true;
        } catch (SQLException e) {
            //TODO Auto-generated catch block
            e.printStackTrace();
        } finally {
            DBConnection.close(rs);
            DBConnection.close(ps);
```

```
            DBConnection.close(con);
        }
        return b;
    }
    //实现修改密码功能
    public boolean upadatePassword(User u) {
        boolean b = false;
        Connection con = DBConnection.getOneCon();
        PreparedStatement ps = null;
        try {
            ps = con.prepareStatement("update usertable set password = ? where userName = ?");
            ps.setString(1, u.getUpass());
            ps.setString(2, u.getUname());
            int i = ps.executeUpdate();
            if (i > 0)
                b = true;
        } catch (SQLException e) {
            //TODO Auto-generated catch block
            e.printStackTrace();
        } finally {
            DBConnection.close(ps);
            DBConnection.close(con);
        }
        return b;
    }
}
```

11.5 系统实现

11.5.1 用户注册

当新用户注册时,该模块要求用户必须输入姓名和密码信息,否则不允许注册。注册成功的用户信息被存入 usertable 表中。

1. 视图(JSP 页面)

该模块中只有一个 JSP 页面:regist.jsp,该页面负责提供注册信息的输入界面,如图 11.4 所示。另外,在注册页面中使用了一个隐藏域 flag,控制器 servlet 根据 flag 值,检查用户名是否存在。

regist.jsp 的核心代码如下:

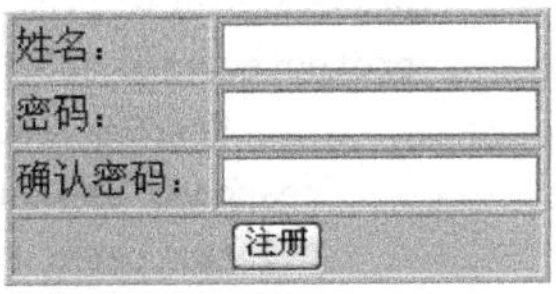

```
<body>
    <form action = "registServlet" method = "post" name = "registForm">
        <input type = "hidden" name = "flag">
        <table border = 1 bgcolor = "lightblue" align = "center">
            <tr>
                <td>姓名:</td>
```

图 11.4 注册页面

```html
                    <td>
                        <input class="textSize" type="text" name="uname" maxlength="20"
                            value="${requestScope.userName}" onblur="nameIsNull()"/>
                        <c:if test="${requestScope.isExit == 'false'}">
                            <font color=red size=5>×</font>
                        </c:if>
                        <c:if test="${requestScope.isExit == 'true'}">
                            <font color=green size=5>√</font>
                        </c:if>
                    </td>
                </tr>
                <tr>
                    <td>密码:</td>
                    <td><input class="textSize" type="password" maxlength="20" name="upass"/>
                    </td>
                </tr>
                <tr>
                    <td>确认密码:</td>
                    <td><input class="textSize" type="password" maxlength="20" name=
                    "reupass"/></td>
                </tr>
                <tr>
                    <td colspan="2" align="center"><input type="button" value="注册"
                        onclick="allIsNull()"/></td>
                </tr>
            </table>
        </form>
    </body>
```

2. 控制器(servlet)

该控制器 servlet 对象的 urlPatterns 是{ "/registServlet" }。控制器获取视图请求后,将视图信息封装在实体模型 User 中。如果获取的请求是检查用户名(flag 等于 0)是否可用,则调用业务模型 UserBusyness 的 nameIsExit 方法执行业务处理。不管用户名是否存在,都返回 regist.jsp 页面告诉用户。如果获取的请求是注册(flag 等于 1),则调用业务模型 UserBusyness 的 regist 方法实现注册。注册成功回到 login.jsp 页面,注册失败回到 regist.jsp 页面。

RegistServlet.java 的核心代码如下:

```java
@WebServlet(name = "registServlet", urlPatterns = { "/registServlet" })
public class RegistServlet extends HttpServlet {
    private static final long serialVersionUID = 1L;
    protected void doGet(HttpServletRequest request,
            HttpServletResponse response) throws ServletException, IOException {
        response.setContentType("text/html;charset=GBK");
        PrintWriter out = response.getWriter();
        //获得页面提交的信息
        String name = request.getParameter("uname");
        String pass = request.getParameter("upass");
        String flag = request.getParameter("flag"); //flag 是一个隐藏域,判断是检查用户还
```

```java
            //是注册功能
            //创建实体模型 User
            User u = new User();
            u.setUname(name);
            u.setUpass(pass);
            //创建业务模型
            UserBusyness ub = new UserBusyness();
            //检查用户是否可用
            if (flag != null && flag.equals("0")) {
                //用户名可用
                if (ub.nameIsExit(name)) {
                    request.setAttribute("isExit", "true");
                } else {//用户名不可用
                    request.setAttribute("isExit", "false");
                }
                //不管用户名可不可用都返回页面
                request.setAttribute("userName", name);
                //返回到注册页面
                RequestDispatcher dis = request.getRequestDispatcher("regist.jsp");
                dis.forward(request, response);
            }
            //实现注册功能
            if (flag != null && flag.equals("1")) {
                if (ub.regist(u)) {
                    //注册成功返回到登录页面
                    out.print("注册成功,3秒后去登录!");
                    response.setHeader("refresh", "3;url = login.jsp");
                } else {
                    //注册失败返回到注册页面
                    out.print("注册失败,请查查原因,3秒后继续注册!");
                    response.setHeader("refresh", "3;url = regist.jsp");
                }
            }
        }
        protected void doPost(HttpServletRequest request,
            HttpServletResponse response) throws ServletException, IOException {
            doGet(request, response);
        }
    }
```

11.5.2 用户登录

用户输入用户名和密码后,系统将对用户名和密码进行验证。如果用户名和密码同时正确,则成功登录,进入系统管理主页面(main.jsp);如果用户名或密码有误,则返回到登录页面继续登录。

1. 视图(JSP 页面)

login.jsp 页面提供登录信息输入功能,效果如图 11.5 所示。

图 11.5 登录界面

login.jsp 的核心代码如下：

```html
<body>
    <form action="loginServlet" method="post" name="loginForm">
        <table border=1 bgcolor="lightblue" align="center">
            <tr>
                <td>姓名：</td>
                <td><input class="textSize" type="text" name="uname" maxlength="20"/></td>
            </tr>
            <tr>
                <td>密码：</td>
                <td><input class="textSize" type="password" name="upass" maxlength="20"/></td>
            </tr>
            <tr>
                <td><input type="button" value="提交" onclick="allIsNull()"/></td>
                <td>没有账号，请<a href="regist.jsp">注册</a>!</td>
            </tr>
        </table>
    </form>
</body>
```

2. 控制器(servlet)

该控制器 servlet 对象的 urlPatterns 是{ "/loginServlet" }。控制器获取视图请求后，将视图信息封装在实体模型 User 中，然后调用业务模型 UserBusyness 中的 login 方法执行登录业务处理。登录成功进入 main.jsp 页面，登录失败返回到 login.jsp 页面。

LoginServlet.java 的核心代码如下：

```java
@WebServlet(name = "loginServlet", urlPatterns = { "/loginServlet" })
public class LoginServlet extends HttpServlet {
    private static final long serialVersionUID = 1L;
    protected void doGet(HttpServletRequest request,
            HttpServletResponse response) throws ServletException, IOException {
        response.setContentType("text/html;charset=GBK");
        PrintWriter out = response.getWriter();
        //获得页面提交的信息
        String name = request.getParameter("uname");
        String pass = request.getParameter("upass");
        //创建实体模型 User
        User u = new User();
        u.setUname(name);
        u.setUpass(pass);
        //创建业务模型
        UserBusyness ub = new UserBusyness();
        //登录成功
        if (ub.login(u)) {
            HttpSession session = request.getSession();
            //登录成功把用户名保存到 session 中
```

```
            session.setAttribute("user", u);
            RequestDispatcher dis = request.getRequestDispatcher("main.jsp");
            dis.forward(request, response);
        } else {//登录失败
            out.print("登录失败,查看用户名和密码是否错误?3秒后继续登录!");
            response.setHeader("refresh", "3;url = login.jsp");
        }
    }
    protected void doPost(HttpServletRequest request,
            HttpServletResponse response) throws ServletException, IOException {
        doGet(request, response);
    }
}
```

11.5.3 添加朋友信息

用户输入朋友姓名、生日、电话、E-mail、照片、地址、关系等信息后,该模块首先检查输入的信息是否合法(比如,日期和 E-mail)。如果合法,则实现添加;如果不合法,则提示信息修改。

1. 视图(JSP 页面)

addFriend.jsp 页面实现添加朋友信息的输入界面。该页面中朋友 ID 由主键产生器自动产生,效果如图 11.6 所示。

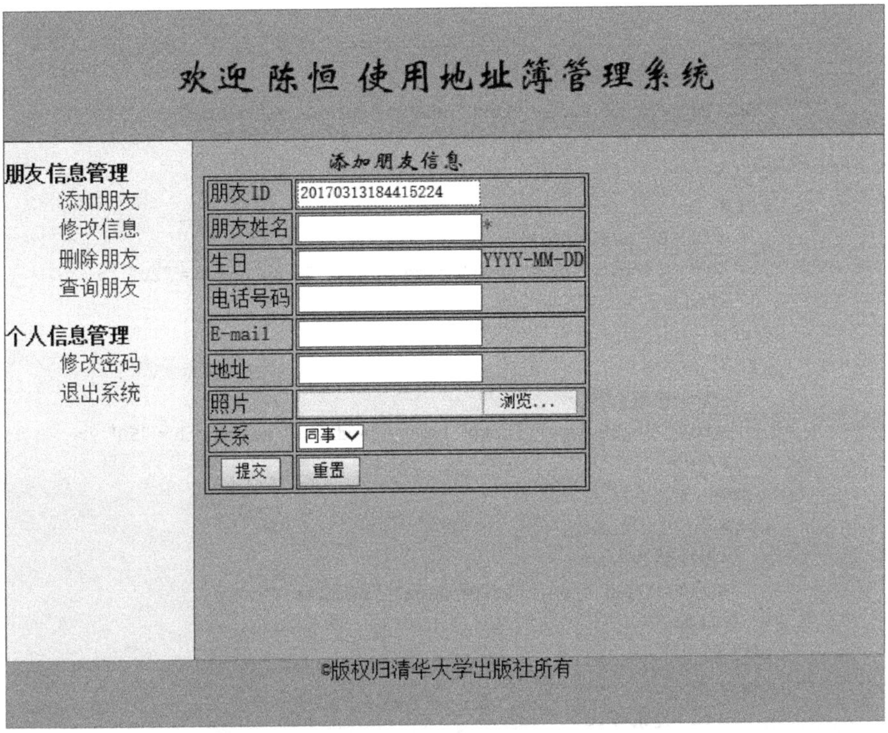

图 11.6 添加朋友

addFriend.jsp 的核心代码如下:

```html
<body>
    <form action = "addFriendServlet" method = "post" name = "addFriendForm" enctype =
    "multipart/form-data">
        <table border = 1>
            <caption>
                <font size = 4 face = 华文新魏>添加朋友信息</font>
            </caption>
            <tr>
                <td>朋友ID</td>
                <td><input type = "text" readonly
                    style = "border-width: 1pt; border-style: dashed; border-color: red"
                    name = "id" value = '<% = MyUtil.getStringID() %>'></td>
            </tr>
            <tr>
                <td>朋友姓名</td>
                <td><input type = "text" name = "name" maxlength = "20" />
                <font color = "red">*</font>
                </td>
            </tr>
            <tr>
                <td>生日</td>
                <td><input type = "text" name = "birthday" maxlength = "10"
                onfocus = "showCalendar(this)" /> YYYY-MM-DD</td>
            </tr>
            <tr>
                <td>电话号码</td>
                <td><input type = "text" name = "telephone" maxlength = "20" />
                </td>
            </tr>
            <tr>
                <td>E-mail</td>
                <td><input type = "text" name = "email" maxlength = "20" />
                </td>
            </tr>
            <tr>
                <td>地址</td>
                <td><input type = "text" name = "address" maxlength = "50" />
                </td>
            </tr>
            <tr>
                <td>照片</td>
                <td><input type = "file" name = "picture"/>
                </td>
            </tr>
            <tr>
                <td>关系</td>
                <td><select name = "relation">
                    <option value = "同事">同事</option>
                    <option value = "同学">同学</option>
```

```html
                <option value="战友">战友</option>
                <option value="老乡">老乡</option>
                <option value="亲戚">亲戚</option>
                <option value="家人">家人</option>
                <option value="密友">密友</option>
                <option value="其他">其他</option>
            </select></td>
        </tr>
        <tr>
            <td align="center"><input type="button" onclick="nameisnull()" value="提交"/>
            </td>
            <td align="left"><input type="reset" value="重置"/>
            </td>
        </tr>
    </table>
</form>
</body>
```

2. 控制器(servlet)

该控制器 servlet 对象的 urlPatterns 是{ "/addFriendServlet" }。控制器获取视图请求后,将视图信息封装在实体模型 Friend 中,然后调用业务模型 FriendBusyness 的 addFriend 方法执行添加业务处理。添加成功进入 queryFriend.jsp 页面,添加失败返回到 addFriend.jsp 页面。

AddFriendServlet.java 的核心代码如下:

```java
@WebServlet(name = "addFriendServlet", urlPatterns = { "/addFriendServlet" })
@MultipartConfig(maxFileSize = 10 * 1024 * 1024)//设置上传文件的最大值为10MB
public class AddFriendServlet extends HttpServlet {
    private static final long serialVersionUID = 1L;
    protected void doGet(HttpServletRequest request,
            HttpServletResponse response) throws ServletException, IOException {
        //获得页面提交的信息
        String id = request.getParameter("id");
        String name = request.getParameter("name");
        String birthday = request.getParameter("birthday");
        String telephone = request.getParameter("telephone");
        String email = request.getParameter("email");
        String address = request.getParameter("address");
        String relation = request.getParameter("relation");
        //创建实体模型
        Friend f = new Friend();
        f.setId(id);
        f.setName(name);
        if (birthday == null || birthday.length() == 0) {
            birthday = "1900-01-01";          //没有输入生日时,设置默认值
        }
        f.setBirthday(birthday);
        if (telephone == null || telephone.length() == 0) {
```

```java
        telephone = "没有电话";              //没有输入电话时,设置默认值
    }
    f.setTelephone(telephone);
    if (email == null || email.length() == 0) {
        email = "没有email";                 //没有输入E-mail时,设置默认值
    }
    f.setEmail(email);
    if (address == null || address.length() == 0) {
        address = "没有地址";                //没有输入地址时,设置默认值
    }
    f.setAddress(address);
    f.setRelation(relation);
    //开始上传照片
    //获得part对象
    Part part = request.getPart("picture");
    //获得原始文件名
    String oldName = MyUtil.getFileName(part);
    //如果选择照片
    if(null != oldName){
        //指定上传的文件保存到服务器的uploadFiles目录中
            File uploadFileDir = new File(getServletContext().getRealPath("/
            uploadFiles"));
            if(!uploadFileDir.exists()){
                uploadFileDir.mkdir();
            }
        //如果传送的文件名是全路径名,取出原始文件名
        int index = oldName.lastIndexOf(File.separator);
        if(index > 0){
            oldName = oldName.substring(index + 1, oldName.length());
        }
        //获取文件类型
        String fileType = oldName.substring(oldName.lastIndexOf("."));
        //以一定规范的文件名上传,因为客户上传时,文件名有可能重名
        String newName = MyUtil.getStringID() + fileType;
        f.setPicture(newName);
        //上传到服务器的uploadFiles目录中
        part.write(uploadFileDir + File.separator + newName);
    }
    //结束上传照片
    HttpSession session = request.getSession();
    //从session中获取用户名
    String userName = ((User) session.getAttribute("user")).getUname();
    //创建业务模型
    FriendBusyness fb = new FriendBusyness();
    //添加成功返回到查询页面
    if (fb.addFriend(f, userName)) {
        RequestDispatcher dis = request.getRequestDispatcher("queryFriendServlet");
        dis.forward(request, response);
    } else {
        //添加失败返回到添加画面
        RequestDispatcher dis = request
```

```
                    .getRequestDispatcher("addFriend.jsp");
            dis.forward(request, response);
        }
    }
    protected void doPost(HttpServletRequest request,
            HttpServletResponse response) throws ServletException, IOException {
        doGet(request, response);
    }
}
```

11.5.4 查询朋友信息

单击系统管理主页面的"查询朋友"链接,打开查询页面 queryFriend.jsp。该链接的目标地址是一个 sevelet,servlet 对象的 urlPatterns 是{ "/queryFriendServlet" }。在该 servlet 中,调用业务模型 FriendBusyness 的 getAllFriends 方法执行查询业务处理,把查询结果显示在查询页面 queryFriend.jsp 中。

1. 控制器(servlet)

QueryFriendServlet.java 的核心代码如下:

```
@WebServlet(name = "queryFriendServlet", urlPatterns = { "/queryFriendServlet" })
public class QueryFriendServlet extends HttpServlet {
    private static final long serialVersionUID = 1L;
    protected void doGet(HttpServletRequest request,
            HttpServletResponse response) throws ServletException, IOException {
        //创建业务模型
        FriendBusyness fb = new FriendBusyness();
        HttpSession session = request.getSession();
        //得到用户名
        String userName = ((User) session.getAttribute("user")).getUname();
        //获得朋友信息
        ArrayList<Friend> al = fb.getAllFriends(userName);
        //把数组放到 request 里传给 JSP 页面
        request.setAttribute("friends", al);
        //获得链接参数
        String flag = request.getParameter("flag");
        //转发到删除页面
        if (flag != null && flag.equals("del")) {
            RequestDispatcher dis = request
                    .getRequestDispatcher("deleteFriend.jsp");
            dis.forward(request, response);
            return;
        }
        //转发到修改页面
        if (flag != null && flag.equals("update")) {
            RequestDispatcher dis = request
                    .getRequestDispatcher("modifyFriend.jsp");
```

```
                dis.forward(request, response);
                return;
            }
            //转发到查询页面
            RequestDispatcher dis = request.getRequestDispatcher("queryFriend.jsp");
            dis.forward(request, response);
        }
        protected void doPost(HttpServletRequest request,
                HttpServletResponse response) throws ServletException, IOException {
            doGet(request, response);
        }
    }
```

2. 视图(JSP 页面)

在该视图中显示成功登录用户的所有朋友信息，如图 11.7 所示。

图 11.7 所有朋友信息

queryFriend.jsp 的核心代码如下：

```
<body>
    <table border = "1">
        <tr bgcolor = "LightGreen">
            <th>朋友 ID</th>
            <th>朋友姓名</th>
            <th>关系</th>
            <th>查看详情</th>
        </tr>
<c:forEach var = "friend" items = "${requestScope.friends}">
    <tr>
        <td>${friend.id}</td>
        <td>${friend.name}</td>
        <td>${friend.relation}</td>
        <td><a href = "freindDetailServlet?op = detail&&friendid = ${friend.id}" target
            = "_blank" style = "text - decoration:none">查看详情</a></td>
    </tr>
</c:forEach>
```

        </table>
    </body>

11.5.5 查看详情

单击图 11.7 中"查看详情"链接,打开详情页面 freindDetail.jsp。该链接的目标地址是一个 sevelet,servlet 对象的 urlPatterns 是{ "/freindDetailServlet" }。在该 servlet 中,首先调用 FriendBusyness 的 getAFriend 方法根据朋友 ID 查询一个朋友信息;然后根据 op 的值(detail)将查询结果显示在详情页面 freindDetail.jsp。详情页面 freindDetail.jsp 的运行效果如图 11.8 所示。

图 11.8 查看详情页面

1. 视图(JSP 页面)

freindDetail.jsp 的核心代码如下:

```
<body bgcolor = "LightCyan">
    <table border = 1>
        <caption><font size = 4 face = 华文新魏>朋友详细信息</font></caption>
        <tr>
            <td>朋友 ID</td>
            <td>${requestScope.friend.id}</td>
        </tr>
        <tr>
            <td>朋友姓名</td>
            <td>${requestScope.friend.name}</td>
        </tr>
        <tr>
            <td>生日</td>
            <td>${requestScope.friend.birthday}</td>
        </tr>
        <tr>
            <td>电话号码</td>
            <td>${requestScope.friend.telephone}</td>
        </tr>
        <tr>
```

```html
            <td>E-mail</td>
            <td>${requestScope.friend.email}</td>
        </tr>
        <tr>
            <td>地址</td>
            <td>${requestScope.friend.address}</td>
        </tr>
        <tr>
            <td>照片</td>
            <td><img src="uploadFiles/${requestScope.friend.picture}" width="100px" height="100px"/></td>
        </tr>
        <tr>
            <td>关系</td>
            <td>${requestScope.friend.relation}</td>
        </tr>
    </table>
</body>
```

2. 控制器(servlet)

FreindDetailServlet.java 的核心代码如下：

```java
@WebServlet(name = "freindDetailServlet", urlPatterns = { "/freindDetailServlet" })
public class FreindDetailServlet extends HttpServlet {
    private static final long serialVersionUID = 1L;
    protected void doGet(HttpServletRequest request,
            HttpServletResponse response) throws ServletException, IOException {
        //创建业务模型
        FriendBusyness fb = new FriendBusyness();
        //获得链接的参数
        String id = request.getParameter("friendid");
        String op = request.getParameter("op");
        //获得朋友信息
        Friend f = fb.getAFriend(id);
        request.setAttribute("friend", f);
        //查看详细信息
        if (op != null && op.equals("detail")) {
            RequestDispatcher dis = request
                    .getRequestDispatcher("freindDetail.jsp");
            dis.forward(request, response);
        }
        //查看修改的信息
        if (op != null && op.equals("update")) {
            RequestDispatcher dis = request
                    .getRequestDispatcher("freindupdate.jsp");
            dis.forward(request, response);
        }
    }
    protected void doPost(HttpServletRequest request,
            HttpServletResponse response) throws ServletException, IOException {
```

```
        doGet(request, response);
    }
}
```

11.5.6 修改朋友信息

单击系统管理主页面中"修改信息"链接,打开修改查询页面 modifyFriend.jsp。该链接的目标地址是一个 sevelet,servlet 对象的 urlPatterns 是{ "/queryFriendServlet" }。在该 servlet 中,根据 flag 值,将查询结果显示在修改查询页面 modifyFriend.jsp。

单击 modifyFriend.jsp 页面中"修改"链接打开修改朋友信息页面 freindupdate.jsp。在页面 freindupdate.jsp 中进行朋友信息修改。单击"修改"按钮,将要修改的朋友信息提交给控制器 servlet,servlet 对象的 urlPatterns 是{ "/updateFriendServlet" }。在该 servlet 中将视图信息封装在实体模型 Friend 中,然后调用业务模型 FriendBusyness 的 upadateFriend 方法执行修改的业务处理。修改成功进入 queryFriend.jsp 页面,修改失败返回到 freindupdate.jsp 页面。

1. 视图(JSP 页面)

该模块的视图共有两个:modifyFriend.jsp 和 freindupdate.jsp。modifyFriend.jsp 页面显示用户可以修改的朋友信息,如图 11.9 所示。freindupdate.jsp 页面提供修改信息输入界面,如图 11.10 所示。

图 11.9 可以修改的朋友信息

modifyFriend.jsp 的核心代码如下:

```
<body>
    <table border = "1">
        <tr bgcolor = "LightGreen">
            <th>朋友 ID</th>
            <th>朋友姓名</th>
            <th>关系</th>
            <th>查看详情</th>
        </tr>
        <c:forEach var = "friend" items = " $ {requestScope.friends}">
            <tr>
                <td>${friend.id}</td>
                <td>${friend.name}</td>
                <td>${friend.relation}</td>
```

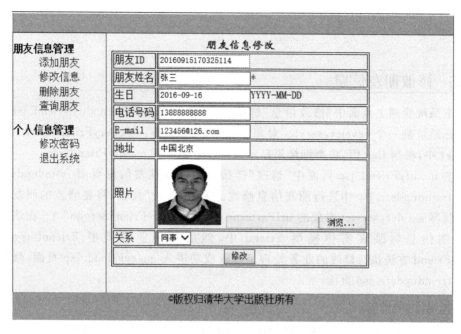

图 11.10 修改信息输入界面

```
            <td align = "center">
                <a href = "freindDetailServlet?op = update&&friendid = ${friend.id}"
                style = "text-decoration: none">修改</a>
            </td>
        </tr>
    </c:forEach>
</table>
</body>
```

freindupdate.jsp 的核心代码如下：

```
<body bgcolor = "LightCyan">
    <form action = "updateFriendServlet" method = "post" name = "updateForm" enctype =
    "multipart/form-data">
    <table border = 1>
        <caption><font size = 4 face = 华文新魏>朋友信息修改</font></caption>
        <tr>
            <td>朋友 ID</td>
            <td><input type = "text" readonly
                style = "border-width:1pt;border-style:dashed;border-color:red"
                name = "id" value = "${requestScope.friend.id}"/></td>
        </tr>

        <tr>
            <td>朋友姓名</td>
            <td><input type = "text" name = "name" maxlength = "20"
                value = "${requestScope.friend.name}"/><font color = "red">*</font></td>
        </tr>
        <tr>
```

```html
         <td>生日</td>
         <td><input type="text" name="birthday" maxlength="10"
         value="${requestScope.friend.birthday}" onfocus="showCalendar(this)"/>
         YYYY-MM-DD</td>
</tr>
<tr>
         <td>电话号码</td>
         <td><input type="text" name="telephone" maxlength="20"
         value="${requestScope.friend.telephone}"/></td>
</tr>
<tr>
         <td>E-mail</td>
         <td><input type="text" name="email" maxlength="20"
         value="${requestScope.friend.email}"/></td>
</tr>
<tr>
         <td>地址</td>
         <td><input type="text" name="address" maxlength="50"
         value="${requestScope.friend.address}"/></td>
</tr>
<tr>
         <td>照片</td>
         <td>
             <img src="uploadFiles/${requestScope.friend.picture}" width=
             "100px" height="100px"/>
             <input type="file" name="picture"/>
         </td>
</tr>
<tr>
         <td>关系</td>
         <td>
             <select name="relation">
                 <option value="同事"
                     <c:if test="${requestScope.friend.relation=='同事'}">selected
                     </c:if>
                     >同事</option>
                 <option value="同学"
                     <c:if test="${requestScope.friend.relation=='同学'}">
                     selected</c:if>
                     >同学</option>
                 <option value="战友"
                     <c:if test="${requestScope.friend.relation=='战友'}">selected
                     </c:if>
                     >战友</option>
                 <option value="老乡"
                     <c:if test="${requestScope.friend.relation=='老乡'}">selected
                     </c:if>
                     >老乡</option>
                 <option value="亲戚"
                     <c:if test="${requestScope.friend.relation=='亲戚'}">selected
                     </c:if>
```

```html
                    >亲戚</option>
                    <option value="家人"
                        <c:if test="${requestScope.friend.relation=='家人'}">selected
                        </c:if>
                        >家人</option>
                    <option value="密友"
                        <c:if test="${requestScope.friend.relation=='密友'}">selected
                        </c:if>
                        >密友</option>
                    <option value="其他"
                        <c:if test="${requestScope.friend.relation=='其他'}">selected
                        </c:if>
                        >其他</option>
                </select>
            </td>
        </tr>
        <tr>
            <td colspan="2" align="center"><input type="button" onclick="nameisnull()"
            value="修改"/></td>
        </tr>
    </table>
</form>
</body>
```

2. 控制器(servlet)

UpdateFriendServlet.java 的核心代码如下：

```java
@WebServlet(name = "updateFriendServlet", urlPatterns = { "/updateFriendServlet" })
@MultipartConfig(maxFileSize = 10 * 1024 * 1024)//设置上传文件的最大值为10MB
public class UpdateFriendServlet extends HttpServlet {
    private static final long serialVersionUID = 1L;
    protected void doGet(HttpServletRequest request,
            HttpServletResponse response) throws ServletException, IOException {
        //获得页面提交的信息
        String id = request.getParameter("id");
        String name = request.getParameter("name");
        String birthday = request.getParameter("birthday");
        String telephone = request.getParameter("telephone");
        String email = request.getParameter("email");
        String address = request.getParameter("address");
        String relation = request.getParameter("relation");
        //创建实体模型
        Friend f = new Friend();
        f.setId(id);
        f.setName(name);
        if (birthday == null || birthday.length() == 0) {
            birthday = "1900-01-01";            //没有输入生日时,设置默认值
        }
        f.setBirthday(birthday);
        if (telephone == null || telephone.length() == 0) {
```

```java
            telephone = "没有电话";              //没有输入电话时,设置默认值
        }
        f.setTelephone(telephone);
        if (email == null || email.length() == 0) {
            email = "没有email";                //没有输入E-mail时,设置默认值
        }
        f.setEmail(email);
        if (address == null || address.length() == 0) {
            address = "没有地址";                //没有输入地址时,设置默认值
        }
        f.setAddress(address);
        f.setRelation(relation);

        //获得part对象
        Part part = request.getPart("picture");
        //获得原始文件名
        String oldName = MyUtil.getFileName(part);
        //如果选择照片
        if(null != oldName){
            //指定上传的文件保存到服务器的uploadFiles目录中
            File uploadFileDir = new File(getServletContext().getRealPath
            ("/uploadFiles"));
            if(!uploadFileDir.exists()){
                uploadFileDir.mkdir();
            }
            //如果传送的文件名是全路径名,取出原始文件名
            int index = oldName.lastIndexOf(File.separator);
            if(index > 0){
                oldName = oldName.substring(index + 1, oldName.length());
            }
            //获取文件类型
            String fileType = oldName.substring(oldName.lastIndexOf("."));
            //以一定规范的文件名上传,因为客户上传时,文件名有可能重名
            String newName = MyUtil.getStringID() + fileType;
            f.setPicture(newName);
            //上传到服务器的uploadFiles目录中
            part.write(uploadFileDir + File.separator + newName);
        }
        //创建业务模型
        FriendBusyness fb = new FriendBusyness();
        if (fb.upadateFriend(f)) {
            RequestDispatcher dis = request.getRequestDispatcher("queryFriendServlet");
            dis.forward(request, response);
        } else {
            //修改失败返回到修改页面
            request.setAttribute("friend", f);      //返回输入的值
            RequestDispatcher dis = request
                    .getRequestDispatcher("freindupdate.jsp");
            dis.forward(request, response);
        }
    }
}
```

```
        protected void doPost(HttpServletRequest request,
                HttpServletResponse response) throws ServletException, IOException {
            doGet(request, response);
        }
    }
```

11.5.7 删除朋友信息

单击系统管理主页面的"删除朋友"链接,打开删除查询页面 deleteFriend.jsp。该链接的目标地址是一个 sevelet,servlet 对象的 urlPatterns 是{ "/queryFriendServlet" }。在该 servlet 中,根据 flag 值,将查询结果显示在删除查询页面 deleteFriend.jsp 中。

在 deleteFriend.jsp 页面中选择要删除的朋友,单击"删除"按钮,将要删除朋友的 ID 提交给控制器 servlet,servlet 对象的 urlPatterns 是{ "/deleteServlet" }。在该 servlet 中调用业务模型 FriendBusyness 的 deleteFriend 方法执行删除的业务处理。成功删除后进入删除查询页面 deleteFriend.jsp。

1. 视图(JSP 页面)

该模块中只有一个视图 deleteFriend.jsp,在该视图中显示用户可以删除的朋友信息,如图 11.11 所示。

图 11.11 可以删除的朋友信息

deleteFriend.jsp 的核心代码如下:

```
<body>
    <form action = "deleteServlet" name = "deleteForm" method = "post">
        <table border = "1">
            <tr bgcolor = "LightGreen">
                <th>朋友 ID </th>
                <th>朋友姓名</th>
                <th>关系</th>
                <th>查看详情</th>
            </tr>
            <c:forEach var = "friend" items = " $ {requestScope.friends}">
            <tr>
                <td><input type = "checkbox" name = "deleteid" value = " $ {friend.id}"/>
                $ {friend.id}</td>
                <td>$ {friend.name}</td>
```

```
                <td>${friend.relation}</td>
                <td><a href = "freindDetailServlet?op = detail&&friendid = ${friend.id}" target
                 = "_blank" style = "text-decoration:none">查看详情</a></td>
            </tr>
        </c:forEach>
        <tr>
            <td colspan = "4" align = "center"><input type = "button" value = "删除" onclick =
             "confirmDelete()"/></td>
        </tr>
    </table>
</form>
</body>
```

2. 控制器(servlet)

DeleteServlet.java 的核心代码如下：

```
@WebServlet(name = "deleteServlet", urlPatterns = { "/deleteServlet" })
public class DeleteServlet extends HttpServlet {
    private static final long serialVersionUID = 1L;
    protected void doGet(HttpServletRequest request,
            HttpServletResponse response) throws ServletException, IOException {
        //创建业务模型
        FriendBusyness fb = new FriendBusyness();
        //获得要删除的 id
        String id[] = request.getParameterValues("deleteid");
        //删除朋友
        fb.deleteFriend(id);
        RequestDispatcher dis = request
                .getRequestDispatcher("queryFriendServlet?flag = del");
        dis.forward(request, response);
    }
    protected void doPost(HttpServletRequest request,
            HttpServletResponse response) throws ServletException, IOException {
        doGet(request, response);
    }
}
```

11.5.8 修改密码

单击系统管理主页面的"修改密码"链接，打开密码修改页面 updatepassword.jsp。在该页面中用户输入新密码，单击"修改"按钮。将密码信息提供给控制器 servlet，servlet 对象的 urlPatterns 是{ "/updatePasswordServlet" }。控制器获取视图请求后，将视图信息封装在实体模型 User 中，然后调用业务模型 UserBusyness 的 updatePassword 方法执行修改密码业务。

1. 视图(JSP 页面)

密码修改页面 updatepassword.jsp 的效果如图 11.12 所示。

图 11.12 密码修改页面

updatepassword.jsp 的核心代码如下:

```html
<body>
    <form action="updatePasswordServlet" method="post" name="updatePasswordForm">
        <table 
        border=1
        bgcolor="lightblue"
        align="center">
            <tr>
                <td>姓名:</td>
                <td>
                    ${sessionScope.user.uname }
                </td>
            </tr>

            <tr>
                <td>新密码:</td>
                <td><input class="textSize" type="password" name="upass"
                value="${sessionScope.user.upass }"/></td>
            </tr>

            <tr>
                <td>确认密码:</td>
                <td><input class="textSize" type="password" maxlength="20"
                name="reupass"/></td>
            </tr>

            <tr>
                <td colspan="2" align="center"><input type="button" value="修改"
                onclick="allIsNull()"/></td>
            </tr>

        </table>
    </form>
</body>
```

2. 控制器(servlet)

UpdatePasswordServlet.java 的核心代码如下:

```java
@WebServlet(name = "updatePasswordServlet", urlPatterns = { "/updatePasswordServlet" })
public class UpdatePasswordServlet extends HttpServlet {
    private static final long serialVersionUID = 1L;
    protected void doGet(HttpServletRequest request, HttpServletResponse response)
            throws ServletException, IOException {
        response.setContentType("text/html;charset=GBK");
        PrintWriter out = response.getWriter();
        //获得页面提交的信息
        String pass = request.getParameter("upass");
        HttpSession session = request.getSession();
        String name = ((User)session.getAttribute("user")).getUname();
        //创建实体模型 User
```

```java
        User u = new User();
        u.setUname(name);
        u.setUpass(pass);
        //创建业务模型
        UserBusyness ub = new UserBusyness();
        //实现修改功能
        if(ub.upadatePassword(u)){
            //注册成功返回到登录页面
            out.print("修改成功,3秒后去登录!");
            response.setHeader("refresh", "3;url = login.jsp");
        }else{
            //失败返回到注册页面
            out.print("修改失败,请查查原因,3秒后继续修改!");
            response.setHeader("refresh", "3;url = upadatepassword.jsp");
        }
    }
    protected void doPost(HttpServletRequest request, HttpServletResponse response)
            throws ServletException, IOException {
        doGet(request,response);
    }
}
```

11.5.9 退出系统

单击系统管理主页面的"退出系统"链接,该链接的目标地址是一个控制器 servlet,servlet 对象的 urlPatterns 是{ "/exitUserServlet" }。在控制器中清空用户会话 session。退出系统成功后,进入登录页面,重新登录。

ExitUserServlet.java 的核心代码如下:

```java
@WebServlet(name = "exitUserServlet", urlPatterns = { "/exitUserServlet" })
public class ExitUserServlet extends HttpServlet {
    private static final long serialVersionUID = 1L;
    protected void doGet(HttpServletRequest request, HttpServletResponse response)
            throws ServletException, IOException {
        response.setContentType("text/html;charset = GBK");
        PrintWriter out = response.getWriter();
        HttpSession session = request.getSession();
        session.removeAttribute("user");
        out.print("退出系统,3秒后重新登录!");
        response.setHeader("refresh", "3;url = login.jsp");

    }
    protected void doPost(HttpServletRequest request, HttpServletResponse response)
            throws ServletException, IOException {
        doGet(request,response);
    }
}
```

参考文献

[1] 耿祥义,张跃平. JSP 实用教程[M]. 2 版. 北京:清华大学出版社,2007.
[2] 孙鑫. Servlet/JSP 深入详解——基于 Tomcat 的 Web 开发[M]. 北京:电子工业出版社,2008.
[3] 赵俊峰,姜宁,焦学理. Java Web 应用开发案例教程——基于 MVC 模式的 JSP+Servlet+JDBC 和 AJAX[M]. 北京:清华大学出版社,2012.
[4] 林信良. JSP & Servlet 学习笔记[M]. 2 版. 北京:清华大学出版社,2012.
[5] 刘俊亮,王清华. JSP Web 开发学习实录[M]. 北京:清华大学出版社,2011.
[6] http://www.w3school.com.cn/.